Springer Optimization and Its Applications

VOLUME 77

Managing Editor
Panos M. Pardalos (University of Florida)

Editor–Combinatorial Optimization
Ding-Zhu Du (University of Texas at Dallas)

Advisory Board
J. Birge (University of Chicago)
C.A. Floudas (Princeton University)
F. Giannessi (University of Pisa)
H.D. Sherali (Virginia Polytechnic and State University)
T. Terlaky (McMaster University)
Y. Ye (Stanford University)

Aims and Scope
Optimization has been expanding in all directions at an astonishing rate during the last few decades. New algorithmic and theoretical techniques have been developed, the diffusion into other disciplines has proceeded at a rapid pace, and our knowledge of all aspects of the field has grown even more profound. At the same time, one of the most striking trends in optimization is the constantly increasing emphasis on the interdisciplinary nature of the field. Optimization has been a basic tool in all areas of applied mathematics, engineering, medicine, economics, and other sciences.

The series *Springer Optimization and Its Applications* publishes undergraduate and graduate textbooks, monographs and state-of-the-art expository work that focus on algorithms for solving optimization problems and also study applications involving such problems. Some of the topics covered include nonlinear optimization (convex and nonconvex), network flow problems, stochastic optimization, optimal control, discrete optimization, multi-objective programming, description of software packages, approximation techniques and heuristic approaches.

For further volumes:
http://www.springer.com/series/7393

Ding-Zhu Du • Peng-Jun Wan

Connected Dominating Set: Theory and Applications

Ding-Zhu Du
Department of Computer Science
University of Texas, Dallas
Richardson, TX, USA

Peng-Jun Wan
Department of Computer Science
Illinois Institute of Technology
Chicago, IL, USA

ISSN 1931-6828
ISBN 978-1-4614-5241-6   ISBN 978-1-4614-5242-3 (eBook)
DOI 10.1007/978-1-4614-5242-3
Springer New York Heidelberg Dordrecht London

Library of Congress Control Number: 2012948679

© Springer Science+Business Media New York 2013
This work is subject to copyright. All rights are reserved by the Publisher, whether the whole or part of the material is concerned, specifically the rights of translation, reprinting, reuse of illustrations, recitation, broadcasting, reproduction on microfilms or in any other physical way, and transmission or information storage and retrieval, electronic adaptation, computer software, or by similar or dissimilar methodology now known or hereafter developed. Exempted from this legal reservation are brief excerpts in connection with reviews or scholarly analysis or material supplied specifically for the purpose of being entered and executed on a computer system, for exclusive use by the purchaser of the work. Duplication of this publication or parts thereof is permitted only under the provisions of the Copyright Law of the Publisher's location, in its current version, and permission for use must always be obtained from Springer. Permissions for use may be obtained through RightsLink at the Copyright Clearance Center. Violations are liable to prosecution under the respective Copyright Law.

The use of general descriptive names, registered names, trademarks, service marks, etc. in this publication does not imply, even in the absence of a specific statement, that such names are exempt from the relevant protective laws and regulations and therefore free for general use.

While the advice and information in this book are believed to be true and accurate at the date of publication, neither the authors nor the editors nor the publisher can accept any legal responsibility for any errors or omissions that may be made. The publisher makes no warranty, express or implied, with respect to the material contained herein.

Printed on acid-free paper

Springer is part of Springer Science+Business Media (www.springer.com)

Preface

As a combinatorial subject, the connected dominating set has been studied in as early as 1970s. However, it was not a major subject and hence did not attract much attention. This situation had changed since 1998 due to its important applications in communication and computer networks, especially its role as a virtual backbone in wireless networks. During the last fourteen years, a large amount of research papers have been published in the theory and applications of the connected dominating set. However, until this book, no attempt has been made to put results on this topic into a collection. When we started to collect references and to make our plan, we realized that actually there exist too much materials in the literature to put into a single book. All materials can be classified into three categories, study in combinatorial structures, applications in networking systems, and theory with computational nature. We decided to give up the first two categories and hence keep this book as a theory-oriented one with computational nature. Therefore, in each topic, we emphasize on theoretical developments on computational complexity and algorithm designs and analysis. Since each theoretical development is motivated from some applications in the real world, we start each chapter, except the first one for introduction, with a section on motivation and overview. This is a reference book which presents the state of art in research from a computational aspect of the connected dominating set. It may also serve as a textbook for advanced topics in a graduate course in applied mathematics, operations research, and computer science. Indeed, we have used some chapters of this book to teach a seminar course (CS7301) in the University of Texas at Dallas, a wireless networking course (CS547) in the Illinois Institute of Technology, and some short courses in graduate summer schools in China. We wish to express our thanks to Hejiao Huang, Hong Zhu, Zhenhua Duan, and Tiende Guo for the organization of such summer courses.

We wish to express our thanks to the colleagues and graduate students, Weili Wu, Zhao Zhang, Wonjun Lee, Deying Li, Yuexuan Wang, Donghyun Kim, Manki Min, Xiaofeng Gao, Feng Zou, Xiuzhen Cheng, Mihaela Cardei, My Thai, Ling Ding, James Willson, Lidong Wu, Zaixin Lu, Kai Xing, Lidan Fan, Wen Xu, Yuanjun Bi, Yuqing Zhu, Huan Ma, Lan Li, Khaled Alzoubi, Ophir Frieder, Xiaohua Xu,

Zhu Wang, and Mining Li, who have made direct or indirect contributions in the process of writing this book.

Especially, we wish to thank Professors Andy Yao, Francis Yao, and Xiaohua Jia for their support. A lot of work on this book was done during our visit at Institute for Interdisciplinary Information Sciences, Tsinghua University, and City University of Hong Kong.

Richardson, TX, USA Ding-Zhu Du
Chicago, IL, USA Peng-Jun Wan

Contents

1 Introduction 1
    1.1   Connected Domination Number 1
    1.2   Virtual Backbone in Wireless Networks 3
    1.3   Converter Placement in Optical Networks 5
    1.4   Connected Domatic Number 6
    1.5   Lifetime of Sensor Networks 8
    1.6   Theory and Applications 9

2 CDS in General Graph 11
    2.1   Motivation and Overview 11
    2.2   Complexity of Approximation 13
    2.3   Two-Stage Greedy Approximation 14
    2.4   Weakly CDS 17
    2.5   One-Stage Greedy Approximation 20
    2.6   Weighted CDS 29
    2.7   Directed CDS 33

3 CDS in Unit Disk Graph 35
    3.1   Motivation and Overview 35
    3.2   NP-Hardness and PTAS 37
    3.3   Two-Stage Algorithm 44
    3.4   Independent Number (I) 47
    3.5   Independent Number (II) 54
    3.6   Zassenhaus–Groemer–Oler Inequality 57

4 CDS in Unit Ball Graphs and Growth Bounded Graphs 63
    4.1   Motivation and Overview 63
    4.2   Gregory–Newton Problem 64
    4.3   Independent Points in Two Balls 67
    4.4   Growth-Bounded Graphs 69
    4.5   PTAS in Growth-Bounded Graphs 73

5	**Weighted CDS in Unit Disk Graph**	77
	5.1 Motivation and Overview	77
	5.2 Node-Weighted Steiner Tree	78
	5.3 Double Partition	80
	5.4 Cell Decomposition	82
	5.5 6-Approximation	86
	5.6 4-Approximation	92
	5.7 3.63-Approximation	96
6	**Coverage**	105
	6.1 Motivation and Overview	105
	6.2 Max-Lifetime Connected Coverage	107
	6.3 Domatic Partition	113
	6.4 Min-Weight Dominating Set	117
7	**Routing-Cost Constrained CDS**	119
	7.1 Motivation and Overview	119
	7.2 Complexity in General Graphs	121
	7.3 CDS with Constraint (ROC1)	124
	7.4 CDS with Constraint (ROCα) for $\alpha \geq 5$	125
8	**CDS in Disk-Containment Graphs**	133
	8.1 Motivation and Overview	133
	8.2 Local Independence Number	134
	8.3 Independence Number	146
	8.4 Greedy Approximation for MIN-CDS	146
9	**CDS in Disk-Intersection Graphs**	151
	9.1 Motivation and Overview	151
	9.2 Voronoi Diagram and Dual of Disks	152
	9.3 Local Search for MIN-DS	155
	9.4 A Two-Staged Algorithm for MIN-CDS	159
10	**Geometric Hitting Set and Disk Cover**	161
	10.1 Motivation and Overview	161
	10.2 Minimum Geometric Hitting Set	161
	10.3 Minimum Disk Cover	165
11	**Minimum-Latency Scheduling**	169
	11.1 Motivation and Overview	169
	11.2 Geometric Preliminaries	171
	11.3 Dominating Tree	174
	11.4 Broadcast Scheduling	177
	11.5 Aggregation Scheduling	178
	11.6 Gathering Scheduling	179
	11.7 Gossiping Scheduling	181

12	**CDS in Planar Graphs**	183
	12.1 Motivation and Overview	183
	12.2 Preliminaries	184
	12.3 Algorithm Description	185
	12.4 Performance Analysis	187

References 193

Index 201

Chapter 1
Introduction

> *Most practical questions can be reduced to problems of largest and smallest magnitudes ... and it is only by solving these problems that we can satisfy the requirements of practice which always seeks the best, the most convenient.*
>
> P. L. ČEBYŠEV

In this chapter, we introduce basic concepts, fundamental results and applications of connected dominating sets.

1.1 Connected Domination Number

Consider a graph $G = (V, E)$. A subset of vertices, D, is called a *dominating set* if every vertex is either in D or adjacent to a vertex in D. If D, in addition, induces a connected subgraph, then it is called a *connected dominating set (CDS)*. The *connected domination number* of a graph G is the minimum cardinality of a CDS, denoted by $\gamma_c(G)$. A CDS that has the size equal to the domination number is called a *minimum CDS*.

The connected domination number is a classical subject studied in graph theory for many years [94]. Some interesting results are obtained in those earlier efforts. The following are two examples.

Let $\ell(G)$ denote the *max leaf number* of a graph G, which is the maximum number of leaves in a spanning tree of G.

Theorem 1.1.1 (Douglas [35]). *For any graph of order n,*

$$\gamma_c(G) = n - \ell(G).$$

Proof. It is easy to see that for any tree T, $\gamma_c(T) = |V(T)| - \ell(T)$. Moreover, a CDS for a spanning tree T of G is also a CDS for G. Therefore, $\gamma_c(G) \leq n - \ell(G)$.

Fig. 1.1 Graph G in the proof of Theorem 1.1.1

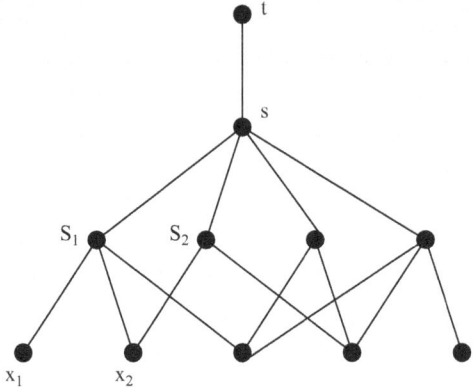

Now, consider a minimum CDS D of G. Let H be a spanning tree of $G[D]$ where $G[D]$ is the subgraph of G induced by D. Connect H to every vertex in $V - D$ to obtain a spanning tree T of G. Then, every vertex in $V - D$ is a leaf of T. Conversely, every leaf of T is in $V - D$. Otherwise, if T has a leaf x not in $T - D$, then $D - \{x\}$ would be a CDS for T and hence a CDS for G, contradicting the minimality of D (Fig. 1.1). □

Theorem 1.1.2 (Sampathkumar and Walikar [94]). *Let G be a graph of order $n \geq 4$. Suppose both graph G and its complement $\bar{G}$ are connected. Then*

$$\gamma_c(G) + \gamma_c(\bar{G}) \leq n(n-3).$$

Proof. Note that a tree has at least two leaves. By Theorem 1.1.1, we have $\gamma_c(G) \leq n - 2$. Moreover, G is connected and hence $n - 1 \leq |E(G)|$. Therefore

$$\gamma_c(G) \leq n - 2 = 2(n-1) - n \leq 2|E(G)| - n.$$

Similarly,

$$\gamma_c(\bar{G}) \leq 2|E(\bar{G})| - n.$$

Thus,

$$\gamma_c(G) + \gamma_c(\bar{G}) \leq 2(|E(G)| + |E(\bar{G})|) - 2n = 2\binom{n}{2} - 2n = n(n-3). \quad \square$$

Laskar and Pfaff [71] showed the NP-hardness of computing the connected domination number or the minimum CDS. Namely, the following problem is NP-hard.

MIN-CDS: Given a graph $G = (V, E)$, find a CDS with minimum cardinality.

Remark. We make a different usage of MIN-CDS from minimum CDS that MIN-CDS is for a problem while the minimum CDS is for a subset of vertices.

Theorem 1.1.3. MIN-CDS *is NP-hard.*

Proof. Consider the following problem.

> SET COVER: Given a collection $\mathcal{C}$ of subsets of a base set X and a positive integer $k \leq |X|$, determine whether $\mathcal{C}$ contains a set cover with cardinality at most k, where a *set cover* is a subcollection $\mathcal{A}$ of $\mathcal{C}$ such that every element of X appears in at least one subset in $\mathcal{A}$.

SET-COVER is a well-known NP-complete problem [57]. We construct a reduction from SET-COVER to MIN-CDS as follows.

For input collection $\mathcal{C}$ and base set X in SET-COVER we first construct a bipartite graph H with $n+m$ vertices labeled by all elements $x_1, u_2, \ldots, x_n$ in X and all subsets $S_1, S_2, \ldots, S_m$ in $\mathcal{C}$. An edge exists between two vertices a and b if and only if $a \in b$ or $b \in a$. Graph G is obtained from H by adding two new vertices s and t and connecting s to t and every S_i for $i = 1, 2, \ldots, m$.

Suppose $\mathcal{C}$ has a set cover $\mathcal{A}$ of at most size k. Then the vertices with labels in $\mathcal{A}$ together with s form a CDS with cardinality at most $k+1$.

Conversely, suppose G has a CDS C of size $k' \leq k+1$. Note that C must contains node s in order to dominating node t or connection t to other vertices in C. Furthermore, we claim that if $a \in C$ for some $a \in X$, then $C - \{a\}$ is still a CDS. In fact, to have a path connecting a and s, there must exist $A \in C$ such that $a \in A$. Thus, a can be dominated by A. Moreover, all vertices dominated by a are also dominated by s. Thus, $C - \{a\}$ is still a CDS. Now, let us denote by C' the CDS obtained from C by deleting t and all elements in X. Then, C' contains s and some vertices labeled by subsets $A_1, A_2, \ldots, A_h$ ($h \leq k'-1$) in $\mathcal{C}$. These h subsets $A_1, A_2, \ldots, A_h$ must cover all elements in X. Therefore, G has a CDS of size at most $k+1$ if and only if $\mathcal{C}$ has a set cover of size at most k. □

While MIN-CDS in general is NP-hard, a lot of earlier efforts were made on design of polynomial-time algorithms for special class of graphs, such as series-parallel graphs [113] and permutation graphs [25].

This situation was changed after applications of CDS were found in wireless networks and optical networks. Since then, the study of CDS is toward application-oriented research. A plenty of issues are involved which generate many research problems in theory.

1.2 Virtual Backbone in Wireless Networks

To keep nodes in a wireless network being able to communicate each other, the network is required to have certain connectivity. Such a task is called *topological control*. Inspired by physical backbone in classical wired networks, the virtual backbone has been introduced to involve in topological control for wireless networks to reduce the utilization of resource. When the wireless network is formulated as a disk graph, the virtual backbone is a CDS of the graph so that all communications between nodes can be executed through the virtual backbone. In fact, the virtual

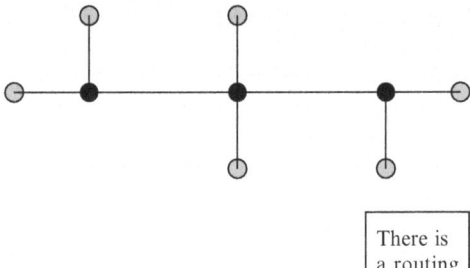

Fig. 1.2 Black nodes form a virtual backbone

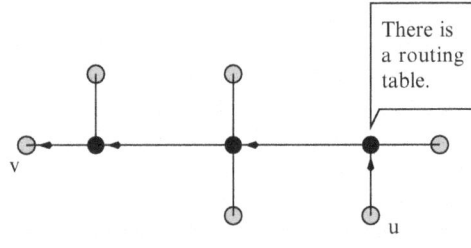

Fig. 1.3 Node u sends a data to node v through a virtual backbone

backbone is required to have two properties: (1) Every node not in virtual backbone should be able directly to communicate with (adjacent to) a node in the virtual backbone. (2) All nodes in the virtual backbone should be able to communicate each other within the virtual backbone, that is, the virtual backbone induces a connected subgraph.

To see the performance of using the virtual backbone, let us consider a network as shown in Fig. 1.2. Note that wireless network has no physical infrastructure. Without using the virtual backbone, every node has to store a routing table in order to be able to communicate with others. With the virtual backbone, only nodes in the virtual backbone need to store a routing table. For nodes not in the virtual backbone, each of them needs to know only an adjacent node in the virtual backbone. In Fig. 1.3, an example is presented to show how node u sends a data to node v through a virtual backbone. Node u first sends the data to its adjacent node w in the virtual backbone C and tells node w that the data is for node v. Since there is a routing table stored at node w, w will figure out a routing path through C to u and then delivers the data along this routing path to node v.

Clearly, we would like to have the virtual backbone as small as possible. This gives a motivation to study MIN-CDS and to design constructions of CDS with small size in various cases [8, 26, 28, 38, 45, 93–95, 98–100, 111, 117–122, 128]. There are several remarkable theoretical results in the literature; each of them made an important progress on the study of CDS as mentioned in the following.

Guha and Khuller [62] designed the first polynomial-time approximation with guaranteed performance ratio $\ln n + O(1)$. They also showed a result on the inapproximability that MIN-CDS cannot have a polynomial-time $\rho \ln n$-approximation for $0 < \rho < 1$ unless $NP \subseteq DTIME(n^{O(\log \log n)})$ where n is the number of vertices in input graph. Improving Guha–Khuller's approximation introduced a study on analysis of greedy approximation with nonsubmodular potential functions. Ruan et al. [92] and Du et al. [40] made a significant contribution in this research direction.

Wan et al. [104] designed the first polynomial-time constant-approximation for MIN-CDS in the unit disk graph which is a mathematical model for homogeneous wireless sensor networks. A plenty of follow-up efforts have been made along this direction.

Cheng et al. [22] designed the first PTAS for MIN-CDS in unit disk graphs, that is a group of polynomial-time $(1+\varepsilon)$-approximation for any $\varepsilon > 0$. This initiates a series of research work on CDS with partition techniques.

Many variations of MIN-CDS or new problems on CDS are proposed recently, motivated from special needs in developments of wireless networking technology. For example Li et al. [75] and Thai et al. [101] proposed the directed CDS; Kim et al. [67] constructed the diameter-bounded CDS; Willson et al. [115] and Ding et al. [32] initiated a study on the routing-cost constrained CDS. Especially, Huang and Gao et al. [53, 66] discovered a technique, double partition, to design better approximations for the weighted CDS problem and related problems. We will study them in later chapters.

1.3 Converter Placement in Optical Networks

One of expectations on next-generation of communication network is to enable people to do remote data gathering and remote scientific experiments. Those applications demand high-speed communication networks with flexible deployment and/or mobile connectivity. One of proposed infrastructures with such properties is the wireless access network on top of the optical core network. Indeed, the optical network in core provides the efficient high-speed communication with high bandwidth and the wireless network in access provides mobile communication or/and flexible deployment. The advantage of fiber-optical backbone network combined with wireless technology has gained more and more interests in the study of the next generation communication network.

An optical network can be considered as a graph $G = (V, E)$ that each edge is associated with a set of wavelengths [60, 73, 81, 82, 88]. The multicast/broadcast/unicast routing requires the existence of a spanning subgraph of G. If a message from an edge to another edge uses different wavelengths, then a converter is required at the common endpoint of the two edges.

Let us use a color to represent a connected component in a subgraph induced by all edges with a certain wavelength. Each converter would connect two connected components into one. To save resource, a minimization problem is formulated [90, 91] as follows.

CONVERTER PLACEMENT: Given a graph $G = (V, E)$ and color-sets for each edge of G such that for every color all edges in the color form a connected subgraph, find the minimum number of vertices such that placing converters on them would connect some colors into a connected spanning subgraph of G.

CONVERTER PLACEMENT can be reduced to MIN-CDS. To do so, we construct another graph G' with vertex set V. Two vertices u and v are connected with an edge if and only if they are in the same color of G. Without loss of generality, we may assume that no color covers all vertices because in such a case, no converter is required. Under the assumption, we can show that a vertex subset C is a feasible solution for CONVERTER-PLACEMENT if and only if C is a CDS in G'.

First, suppose C is a feasible solution. Then every vertex x must be adjacent to a converter; otherwise, it cannot communicate with any converter which is also a vertex in G. Moreover, C must induce a connected subgraph in G' since, otherwise, two converters in different connected components cannot communicate each other. Therefore, C is a CDS in G'.

Conversely, if C is a CDS in G', then there is a spanning tree T of G' with all internal vertices in C. Thus, placing converters at all vertices in C would connect all colors appearing in T together, which is clearly covering T. Therefore, C is a feasible solution for CONVERTER PLACEMENT.

In optical networks, there is also an amplifier placement problem related to CDS. In fact, an optical network usually consists of passive optical star couplers as nodes which are linked with undirectional fibers. When a signal is traveling too long or splits at some couplers, its power may become too weak and hence it needs an amplifier to increase its power to certain level. The minimization of number of amplifiers under certain network connectivity constraint can also be reduced to a special case of MIN-CDS [87].

1.4 Connected Domatic Number

The *domatic number* of a graph G, $\kappa(G)$, is the maximum number of disjoint dominating sets in G. The *connected domatic number* of a graph G, $\kappa_c(G)$, is the maximum number of disjoint CDS in G. $\kappa(G)$ and $\kappa_c(G)$ are very different. The following theorem indicates this fact.

Theorem 1.4.1. *For any positive integer k, there exists a graph G such that*

$$\kappa(G) - \kappa_c(G) = k.$$

Proof. Let $K = (V, E)$ and $K' = (V', E')$ be two disjoint complete graphs of order $k + 1$. Add an edge (v, v') between K and K' for a vertex $v \in V$ and a vertex $v' \in V'$. Denote by G the resulting graph. Then $\kappa(G) = k + 1$ and $\kappa_c(G) = 1$. □

Computing $\kappa_c(G)$ is equivalent to the following problem.

MAX#CDS: Given a graph $G = (V, E)$, find the maximum number of disjoint CDS's.

This problem is also intractable.

Theorem 1.4.2 (Cardei et al. [16]). MAX#CDS *is NP-hard.*

Proof. The NP-completeness of the following problem is proved in [16].

1.4 Connected Domatic Number

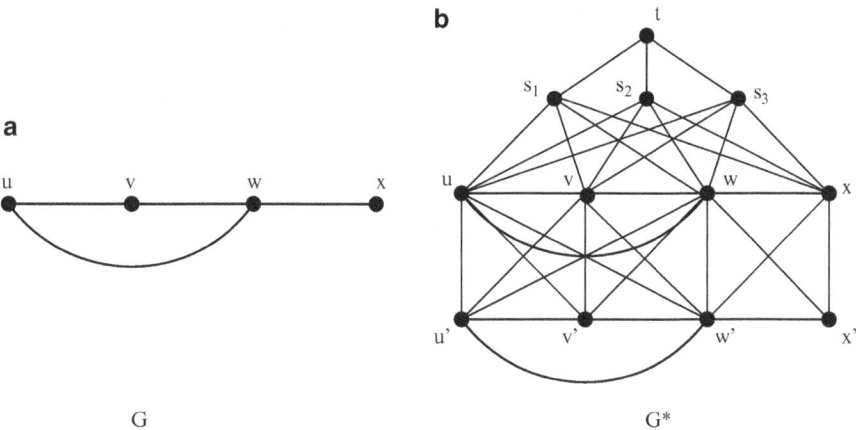

Fig. 1.4 Reduction in the proof of Theorem 1.4.2

3DDS: Given a graph $G = (V,E)$, determine whether G contains three disjoint dominating sets.

We now construct a polynomial-time reduction from 3DDS to MAX#CDS. For input graph $G = (V,E)$ of 3DDS, we make a copy of V, $V' = \{v' \mid v \in V\}$. Connect each vertex $u \in V$ to u' and v' for all $(u,v) \in E$. Add four new vertices s_1, s_2, s_3 and t. Connect every s_i to t for $i = 1,2,3$ and connect every $u \in V$ to every s_i for $i = 1,2,3$. Let $G^* = (V^*, E^*)$ be the graph obtained from the above construction (Fig. 1.4), that is,

$$V^* = V \cup V' \cup \{s_1, s_2, s_3, t\}$$
$$E^* = E \cup \{(u,v') \mid (u,v) \in E\} \cup \{(u,s_i) \mid u \in V, 1 \le i \le 3\}$$
$$\cup \{(s_i, t) \mid 1 \le i \le 3\} \cup \{(u,u') \mid u \in V\}.$$

We show that G contains three disjoint dominating sets if and only if G^* contains three disjoint CDS's. To do so, we first assume that G contains three disjoint dominating sets D_1, D_2, D_3. Then $D_1 \cup \{s_1\}$, $D_2 \cup \{s_2\}$ and $D_3 \cup \{s_3\}$ are three disjoint CDS's for G^*.

Conversely, assume G^* contains three disjoint CDS's C_1, C_2 and C_3. Define $D_i = C_i \cap V$ for $i = 1,2,3$. Then D_1, D_2 and D_3 are disjoint. We claim that each D_i is a dominating set in G. In fact, if there exist a vertex $v \in V$ which cannot be dominated by D_i, then v' cannot be dominated by C_i because every vertex $v' \in V'$ can be dominated by only some vertices in V and v' is dominated by $u \in V$ if and only if v is dominated by u.

Since above construction is done clearly in polynomial-time, MAX#CDS is NP-hard. □

1.5 Lifetime of Sensor Networks

When a very large number of sensors are randomly deployed in target field, the existence of redundant sensors implies the existence of disjoint CDS's. By properly scheduling activation/sleep time of sensors, those CDS's can be organized working in different period as virtual backbone so that the lifetime of the sensor networks is equal to the lifetime of a sensor multiplying the number of disjoint CDS's. Therefore, the maximization of the number of disjoint CDS's has impact in the lifetime maximization of sensor network. This gives an application of MAX#CDS. Actually, from study on the lifetime of sensor networks, more research problems on CDS have been promoted. The following are some of them.

An improvement of lifetime can be seen from the following example as shown in Fig. 1.5. The graph in Fig. 1.5 does not contain two disjoint CDS's. However, if we organize sensors working in the following way, then the lifetime of sensor network can reach 1.5 when every sensor is supposed to has lifetime one.

1. At the 1st 0.5 time period, CDS $\{v_1, v_2\}$ is active.
2. At the 2nd 0.5 time period, CDS $\{v_2, v_3\}$ is active.
3. At the 3rd 0.5 time period, CDS $\{v_3, v_1\}$ is active.

Motivated from this example, we may study the following problem [129].

CDS-SCHEDULING: Given a graph $G = (V, E)$ and a positive vector $b : V \to R^+$, find a sequence of pairs (C_1, t_1), (C_2, t_2), ..., (C_k, t_k) where $1 \leq k \leq |V|$ to maximize $t_1 + t_2 + \cdots + t_k$ under constraint that $\sum_{i: u \in C_i} t_i \leq b(u)$ for every $u \in V$.

Although our scheduling mechanism is able to make the system lifetime longer, the control complexity is increased. Note that different orderings of those CDS $C_1, .., C_p$ give different control complexities. For example, suppose scheduling is in ordering of

$$C_1 = \{v_1, v_2\}, \quad C_2 = \{v_3, v_4\}, \quad C_3 = \{v_2, v_3\}.$$

Then, sensor v_2 should be activated twice. However, in ordering of C_1, C_3, C_2, none of sensors needs to activate twice. This fact raised a research problem on CDS permutation.

An interesting fact discovered in [5] is that putting a sensor alternatively in active and sleep modes in a proper way may double its lifetime since the battery could be

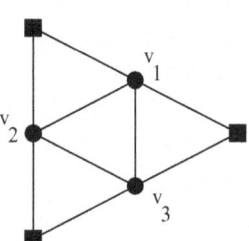

Fig. 1.5 An example

recovered in certain level during sleeping. This fact indicates that CDS permutation contains interesting issues. A proper number of changing between sleep and active modes is good to the lifetime. We may need to find a way to balance the control complexity and the lifetime.

If we allow partial domination, then the lifetime of the system can certainly be increased. A *partial CDS* with percentage p ($0 < p < 1$) is a vertex subset C which dominates at least pn vertices and induces a connected subgraph. There are three types of problems on partial CDS [19, 20].

In the first type of problems, similar to MAX#CDS and CDS SCHEDULING, we want to maximum the lifetime of the system under constraint that the dominating percentage is always kept at least p.

In the second type of problems, the lifetime of network is given, we want to find a sequence of disjoint (or nondisjoint) partial CDS to maximize the minimum dominating percentage p.

In the third type of problems, the lifetime of network is also given, we want to find a sequence of disjoint (or nondisjoint) partial CDS to maximize the sum of products of dominating percentage p and working time of each partial CDS.

1.6 Theory and Applications

In previous sections, we have seen that two mathematical problems MIN-CDS and MAX#CDS have important applications in network technology. Moreover, motivated from those applications, many new mathematical problems and new issues about CDS have been proposed and studied. Especially, as wireless networks and optical networks are developing rapidly, theory of CDS is growing quickly. The aim of this book is to put together recent results on theory and applications of CDS in order to provide the state of arts in this research area for students, professors, researchers in applied mathematics, operations research and computer science.

In each chapter, we first give a motivation and overview, as well as existing open problems, for subject which is going to be studied in the chapter. Then we present theoretical developments. For convenience of the reader, we try to have this book almost self-contained and each chapter also almost self-contained. Therefore, the definition of notations may be defined repeatedly in different chapters.

Also for convenience of the reader, we restrict usage of brief names. Indeed, except DS (dominating set), CDS (connected dominating set), SCDS (strongly connected dominating set), WCDS (weakly connected dominating set), and names of problems, we rarely use brief names for others.

Although most of contents of this book come from research with motivations from applications in the real world, we have to admit that this is a theory book or mathematical book. Therefore, we do not put any computer experimental result in this book.

We wish that this book can be a useful tool in further developments on theory and applications of CDS to enrich contents of combinatorial optimization and computer and communication networks.

Chapter 2
CDS in General Graph

> *Leadership is based on inspiration,*
> *not domination; on cooperation, not intimidation.*
> WILLIAM ATHUR WARD

2.1 Motivation and Overview

Since MIN-CDS is NP-hard, approximation algorithm design becomes an important issue in study of CDS. What is the complexity of approximation for MIN-CDS? Guha and Khuller [62] showed that MIN-CDS has no polynomial-time ($\rho \ln n$)-approximation for $0 < \rho < 1$ unless $NP \subseteq DTIME(n^{O(\log \log n)})$ where n is the number of vertices in input graph. Moreover, they designed a 2-stage greedy algorithm with performance ratio $3 + \ln \delta$ where δ is the maximum vertex degree of input graph. The effort on improvement of this 2-stage greedy algorithm encounted an essential difficulty on analysis of greed approximation.

In 1982, Wolsey [116] discovered a general theorem on analysis of greedy approximation with submodular potential functions, which covers many existing results. For example, the greedy algorithm for WCDS in [17] has a submodular potential function and can be analyzed with Wolsey Theorem. Since Wolsey Theorem was established, the submodularity becomes an important property for algorithm designer to seek. Unfortunately, the potential function used in Guha–Khuller's Greedy Algorithm is not submodular, and so far, no one has found a submodular potential function to design a greedy approximation for MIN-CDS.

How do we analyze the greedy approximation with a nonsubmodular potential function? Ruan et al. [92] found a technique and designed a one-stage greedy approximation for MIN-CDS with performance ratio $2 + \ln \delta$. Du et al. [40] found more techniques and designed a greedy approximation scheme for MIN-CDS with performance ratio $a(1 + \ln \delta)$ for any $a > 1$.

However, those techniques do not work in weighted version of MIN-CDS.

Fig. 2.1 Strongly connected dominating set

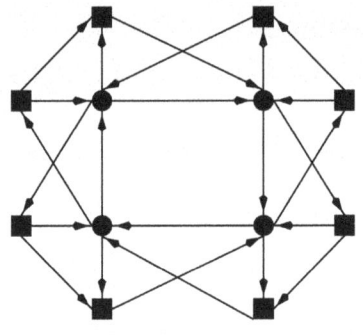

● nodes in strongly connected dominating set

MINW-CDS: Given a connected graph $G = (V,E)$ with vertex weight $w : V \to R^+$, find a CDS with minimum total weight.

For MINW-CDS, Guha and Khuller [62] proposed a two-stage approximation algorithm as follows: At the first stage, construct a dominating set D with a greedy approximation for MINW-SET-COVER in which each node v corresponds to the subset of nodes dominated by the node v.

MINW-SET-COVER: Given a collection C of subsets of X with a nonnegative weight function $w : C \to R^+$, find a set cover with minimum total weight.

At the second stage, connect D into a CDS with a greed approximation, given by Klein and Ravi [69], for node-weighted Steiner tree problem.

NODE-WEIGHTED STEINER TREE: Given a graph $G = (V,E)$ with nonnegative node weight $w : V \to R^+$, and a subset P of nodes, find a subset S of nodes with minimum total weight, interconnecting all nodes in P.

Note that the total weight of dominating D can be upper-bounded by $(1 + \ln n) \cdot$ opt and the total weight of Steiner nodes S added in the second stage can be upper-bounded by $(1 + 2\ln 2) \cdot$ opt where opt is the minimum weight of a CDS. Therefore, the approximation given by Guha and Khuller [62] has performance ratio at most $2 + 3\ln n$. Soon later, Guha and Khuller [63] found that the technique initiated by Klein and Ravi [69] can be directly employed to design greedy approximations for MIN WEIGHT CDS. Actually, this technique works in a wide range of area. The disadvantage of this technique is that obtained performance ratio is a little large. Guha and Khuller [63] also improved this technique. Using improved technique, they designed a polynomial-time approximation for MINW-CDS with performance ratio approaching $1.35 \cdot \ln \delta$, which is the best known so far.

Consider a directed graph $G = (V,E)$. Thai and Du [101] and Li et al. [74] introduce the concept of CDS into directed graphs. A node subset C is a *dominating set* if every node not in C has an arc going to C and an arc coming from C. Furthermore, C is called a *strongly connected dominating set (SCDS)* if subgraph induced by C is strongly connected (Fig. 2.1). They study the following problem.

MIN-SCDS: Given a directed graph, find a strongly connected dominating set with minimum cardinality.

Li et al. [76,77] found a construction of SCDS by using the solution for MINW-BROADCAST and hence obtained polynomial-time $(2+4\ln n)$-approximation [76] and $(2+3\ln n)$-approximation [77] for MIN-SCDS, respectively.

2.2 Complexity of Approximation

Consider the following problem.

MIN-SET-COVER: Given a collection $\mathcal{C}$ of subsets of a finite set X, find a set cover from $\mathcal{C}$, with minimum cardinality.

MIN-SET-COVER has the following inapproximability.

Theorem 2.2.1 (Feige [47]). *For $0 < \rho < 1$, there is no polynomial-time $(\rho \ln n)$-approximation for MIN-SET-COVER unless $NP \subseteq DTIME(n^{O(\log \log n)})$.*

Using this result, Guha and Khuller [62] established the inapproximability of MIN-CDS.

Theorem 2.2.2 (Guha and Khuller [62]). *For $0 < \rho < 1$, MIN-CDS has no polynomial-time $(\rho \ln n)$-approximation unless $NP \subseteq DTIME(n^{O(\log \log n)})$ where n is the number of vertices in input graph.*

Proof. We recall the reduction from SET-COVER to MIN-CDS in the proof of NP-hardness of MIN-CDS in Theorem 1.1.3. The reduction can also be seen as a reduction from MIN-SET-COVER to MIN-CDS as follows.

For any instance of MIN-SET-COVER, a collection $\mathcal{C}$ of m subsets of a set X of n elements, the reduction constructs a graph $G = (V,E)$ with vertex set

$$V = X \cup \mathcal{C} \cup \{s,t\}$$

and edge set

$$E = \{(x,S) \mid x \in S \text{ for } x \in X, S \in \mathcal{C}\} \cup \{(s,S) \mid S \in \mathcal{C}\} \cup \{(s,t)\}.$$

This reduction has been proved to have property that $\mathcal{C}$ has a set cover of size at most k if and only if G has a CDS of size at most $k+1$. Consequently, the minimum set cover of $\mathcal{C}$ contains k subsets if and only if the minimum CDS of G contains $k+1$ vertices.

Now, suppose for some $0 < \rho < 1$, there is a polynomial-time $(\rho \ln n)$-approximation for MIN-CDS. We prove $NP \subseteq DTIME(n^{O(\log \log n)})$.

Choose a positive integer $k_0 > \frac{\rho}{1-\rho}$. Then $\rho(1 + \frac{1}{k_0}) < 1$. Choose a positive number ρ' such that $\rho(1 + \frac{1}{k_0}) < \rho' < 1$. We then show that MIN-SET-COVER in the special case $|\mathcal{C}| = m \leq n$ also has a polynomial-time approximation with performance ratio $\rho' \ln n$.

For each input collection $\mathcal{C}$ in MIN-SET-COVER, we first check all subcollections of at most k_0 subsets whether it is a set cover or not. This takes time bounded by a polynomial of degree k_0.

If no set cover of cardinality k_0 is found, then any set cover of $\mathcal{C}$ contains at least $k_0 + 1$ subsets.

Suppose C is $(\rho \ln n)$-approximation solution for MIN-CDS. Then C has size at most $(\rho \ln(m+n+2))(k+1)$. Thus, we can obtain a set cover $\mathcal{A}$ of size at most $\rho \ln(m+n+2)(k+1) - 1 < \rho(1+\frac{1}{k_0})(1+\frac{\ln 3}{\ln n})(\ln n)k$ where $\mathcal{A} = \mathcal{C} \cap C$. When n is sufficiently large, $\mathcal{A}$ is a $\rho' \ln n$-approximation solution for MIN-SET-COVER. By Theorem 2.2.1, $NP \subseteq DTIME(n^{O(\log \log n)})$. □

There is another lower bound result for MIN-SET-COVER.

Theorem 2.2.3 (Raz and Safra [89]). *There is a constant $c > 0$ such that the existence of polynomial-time $(c \ln n)$-approximation for* MIN-SET-COVER *implies $NP = P$.*

Following from this result, we can also obtain a similar result for MIN-CDS.

Theorem 2.2.4. *There is a constant $c > 0$ such that the existence of polynomial-time $(c \ln n)$-approximation for* MIN-CDS *implies $NP = P$.*

2.3 Two-Stage Greedy Approximation

Consider a graph G and a subset C of vertices in G. We divide all vertices in G into three classes with respect to C:

- *Black* vertices: vertices in C.
- *Grey* vertices: vertices not in C but dominated by black vertices.
- *White* vertices: vertices not dominated by black vertices.

Clearly, C is a CDS if and only if there does not exist a white vertex and the subgraph induced by black vertices is connected. Let $p(C)$ be the number of connected components of $G[C]$, the subgraph of G induced by C, and $h(C)$ the number of white vertices. Let $g(C) = p(C) + h(C)$. Then C is a CDS if and only if $g(C) = 1$. We may use g to design an algorithm as follows:

Greedy Algorithm GK:
input a connected graph G.
Set $C \leftarrow \emptyset$;
while there exists a vertex x such that $g(C \cup \{x\}) < g(C)$ **do**
    choose a vertex x to minimize $g(C \cup \{x\})$ and
    set $C \leftarrow C \cup \{x\}$;
output C.

However, this algorithm may not output a CDS. Indeed, even if for vertex x, $g(C \cup \{x\}) = g(C)$, C may not be a CDS. An example is shown in Fig. 2.2. In fact,

2.3 Two-Stage Greedy Approximation

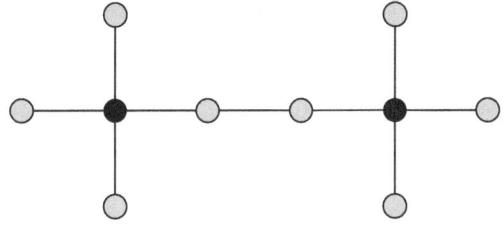

Fig. 2.2 Output of Algorithm GK may not be a CDS

what appeared in this example is a typic case. If a white vertex exists, then let x be a gray vertex adjacent to a white vertex, then we must have $g(C \cup \{x\}) < g(C)$. Therefore, for C obtained from **Greedy Algorithm GK**, no white vertex exists. This means that if output C is not a CDS, then C does not induced a connected subgraph. In such a case, its connected components are apart not very far. Since the given graph is connected, all black components are connected together through some chains of two adjacent gray vertices. To see this, we first note that no gray vertex is adjacent to two black components since coloring such a gray vertex in black would reduce the value of potential function. Now, for contradiction, suppose that all black components cannot be connected through chains of two adjacent gray vertices. Then, we can divide all black vertices into two parts such that the distance between the two parts is more than three, say $k > 3$. Consider the path between the two parts, $(u, x_1, x_2, \ldots, x_{k-1}, v)$, which reaches the distance between the two parts, that is, u and v belong to the two parts respectively, $x_1, x_2, \ldots, x_{k-1}$ are gray vertices with $k - 1 \geq 3$, and no shorter path of this type exists. Since x_2 is grey, it must be adjacent to a black vertex w. If w and u are in the same part, then the path from w to v indicates that the distance between the two parts is at most $k - 1$, a contradiction. If w and v are in the same part, then the path from u to w indicates that the distance between the two parts is at most $3 < k$, also a contradiction.

Based on above observations, Guha and Khuller [62] designed a two-stage greedy algorithm as follows.

Guha–Khuller Algorithm:
input a connected graph G.
Stage 1
      Employ **Greedy Algorithm GK** to obtain a dominating set C;
Stage 2
      while there are more than one black components **do**
           find a chain of two gray vertices x and y connecting at least
           two black components and $C \leftarrow C \cup \{x, y\}$;
output C.

In this two-stage greedy approximation, stage 1 is a greedy algorithm computing a dominating set and stage 2 connects this dominating set into a connected one. In the potential function $g(C)$, $h(C)$ is used for issuing that Stage 1 gives a dominating set, and $p(C)$ is used for making the number of black connected components smaller.

Theorem 2.3.1 (Guha and Khuller [62]). *Suppose input graph is not a star. Then, Guha–Khuller Algorithm is a polynomial-time $(3 + \ln \delta)$-approximation for CDS where δ is the maximum vertex degree of input graph.*

Proof. By a piece, we mean a white vertex or connected component of subgraph induced by black vertices. A piece is said to be *touched* by a vertex x if x is in the piece or adjacent to the piece. For any vertex subset C, the number of piece is $g(C)$. Suppose $x_1, \ldots, x_g$ are selected in turn by Guha–Khuller Algorithm at stage 1. Denote $C_i = \{x_1, \ldots, x_i\}$ for $1 \leq i \leq g$ and $C_0 = \emptyset$. Then, each vertex x touches $1 + g(C_{i-1}) - g(C_{i-1} \cup \{x\})$ pieces with respect to C_{i-1} and x_i reaches the maximum of this number. Suppose opt is the number of vertices in a minimum CDS. Since a dominating set must touch all pieces, there exists a vertex touches at least $\lceil g(C_{i-1})/\text{opt} \rceil$ pieces. Therefore

$$1 + g(C_{i-1}) - g(C_i) \geq \frac{g(C_{i-1})}{\text{opt}}$$

that is,

$$g(C_i) \leq g(C_{i-1})\left(1 - \frac{1}{\text{opt}}\right) + 1.$$

Set $a_i = g(C_i) - \text{opt}$. Then,

$$a_i \leq a_{i-1}\left(1 - \frac{1}{\text{opt}}\right).$$

Clearly, as long as $a_{i-1} > 0$, we have $a_i < a_{i-1}$. Therefore, we must have $a_g \leq 0$. Choose $j \leq g$ such that $a_j \leq 0 < a_{j-1}$. Then $a_g \leq j - g$. This means that when stage 1 ends, at most $\text{opt} - (g - j) + 1$ pieces exist and hence at most $\text{opt} - (g - j) + 1$ connected black components exist. Therefore, at most $2(\text{opt} - g + j)$ vertices would be added in stage 2. Choose i such that $a_{i+1} < \text{opt} \leq a_i$. Then, $j - i \leq \text{opt}$ and

$$\text{opt} \leq a_{i-1}\left(1 - \frac{1}{\text{opt}}\right) \leq a_0\left(1 - \frac{1}{\text{opt}}\right)^i \leq n e^{-i/\text{opt}},$$

where n is the number of vertices of input graph. Thus,

$$i \leq \text{opt} \ln(n/\text{opt}).$$

Therefore,

$$g + 2(\text{opt} - g + j) \leq 2\text{opt} + j \leq 3\text{opt} + i \leq \text{opt}(3 + \ln(n/\text{opt})) \leq \text{opt}(3 + \ln \delta)$$

for $\text{opt} \geq 2$. □

2.4 Weakly CDS

Improving Guha and Khuller's Greedy Algorithm is not an easy job. Actually, this encounted a fundamental difficulty on analysis of greedy algorithms. To explain this, let us use WCDS as an example to introduce the theory of submodular function.

Consider a graph $G = (V,E)$. For any vertex subset C, denote by $q(C)$ the number of connected components of the subgraph with vertex set V and the edge set consisting of all edges incident to vertices in C.

A dominating set C is called a *weakly CDS (WCDS)* if $q(C) = 1$. Chen and Liestman [17] studied the following problem.

MIN-WCDS: Given a graph G, find a WCDS with the minimum cardinality.

They designed a greedy algorithm with potential function $q(C)$.

Chen–Liestman Algorithm
input graph $G = (V,E)$.
$C \leftarrow \emptyset$;
while $q(C) \geq 2$ **do**
    choose $u \in V$ to mimimize $q(C \cup \{u\})$
    $C \leftarrow C \cup \{u\}$;
output C.

The performance ratio of this algorithm is guaranteed by the following theorem.

Theorem 2.4.1 (Chen and Liestman [17]). *Chen–Liestman Algorithm produces an approximation solution within a factor of $(1 + \ln \delta)$ from optimal, where δ is the maximum vertex degree of input graph.*

Actually, the above result has been covered by a general theory on submodular function proposed by Wolsey [116].

Consider a finite set X and a real function f defined on 2^X, the collection of all subsets of X. f is *submodular* if for any two subsets A and B of X,

$$f(A) + f(B) \geq f(A \cup B) + f(A \cap B).$$

f is *increasing* if for $A \subset B$, $f(A) \leq f(B)$. The *marginal value* of B with respect to A is defined by

$$\Delta_B f(A) = f(A \cup B) - f(A).$$

When $B = \{x\}$ for some $x \in X$, we simply write $\Delta_x f(A)$ instead of $\Delta_{\{x\}} f(A)$ for the marginal value of $\{x\}$ (or simply the marginal value of x) with respect to A. Both monotonicity and submodularity of a function f can be characterized in terms of the marginal values [4, 52, 84, 116].

Lemma 2.4.2. *f is submodular and increasing if and only if for any $x \in X$,*

$$A \subset B \Rightarrow \Delta_x f(A) \geq \Delta_x f(B).$$

Wolsey [116] studied the following problem and greedy algorithm.

MIN-SUBMODULAR-COVER: Given a submodular and increasing function $f : 2^X \to R$ and a nonnegative cost function $c : X \to R^+$, find $A \subseteq X$ to minimize $C(A) = \sum_{x \in A} c(x)$ under constraint $f(A) = f(X)$.

Wolsey Greedy Algorithm
input a monotone increasing submodular function $f : 2^X \to R$;
Initially, set $A \leftarrow \emptyset$;
while $f(A) < f(X)$ **do**
$\quad$ choose $x \in X - A$ to maximize $\frac{\Delta_x f(A)}{c(x)}$
$\quad A \leftarrow A \cup \{x\}$;
output A.

The performance of this algorithm is guaranteed by the following theorem [116].

Theorem 2.4.3 (Wolsey Theorem). *Suppose f is a submodular, monotone increasing integer function on 2^X with $f(\emptyset) = 0$. Then, Wolsey Greedy Algorithm produces a $H(\gamma)$-approximation for* MIN-SUBMODULAR-COVER *where $\gamma = \max_{x \in X} f(\{x\})$.*

Proof. Let $x_1, x_2, \ldots, x_k$ be the sequence of elements selected by Wolsey Greedy Algorithm and $A = \{x_1, x_2, \ldots, x_k\}$. Let A^* be an optimal solution of MIN-SUBMODULAR-COVER. We prove

$$c(A) \leq H(\gamma) c(A^*)$$

by a charging argument. Denote $A_0 = \emptyset$ and $A_i = \{x_1, x_2, \ldots, x_i\}$ for each $1 \leq i \leq k$. Denote $\mu_0 = 0$ and $\mu_i = \frac{c(x_i)}{\Delta_{x_i} f(A_{i-1})}$ for each $1 \leq i \leq k$. The parameter μ_i is the referred to as the average price per increment of coverage by x_i for each $1 \leq i \leq k$. We claim that

$$\mu_0 \leq \mu_1 \leq \mu_2 \leq \cdots \leq \mu_k.$$

Indeed, the first inequality is trivial. For any $1 \leq i < k$,

$$\mu_i = \frac{c(x_i)}{\Delta_{x_i} f(A_{i-1})} \leq \frac{c(x_{i+1})}{\Delta_{x_{i+1}} f(A_{i-1})} \leq \frac{c(x_{i+1})}{\Delta_{x_{i+1}} f(A_i)} = \mu_{i+1},$$

where the first inequality follows from the greedy rule and the second inequality follows from the submodularity of f. Thus, our claim holds. Now for iteration i with $1 \leq i \leq k$, we charge each $e \in A^*$ with $\mu_i(\Delta_e f(A_{i-1}) - \Delta_e f(A_i))$. Then, the total charge on each $e \in A^*$ is

$$\sum_{i=1}^{k} \mu_i (\Delta_e f(A_{i-1}) - \Delta_e f(A_i)),$$

2.4 Weakly CDS

and the total charge on A^* is

$$\sum_{e \in S} \sum_{i=1}^{k} \mu_i \left(\Delta_e f(A_{i-1}) - \Delta_e f(A_i) \right).$$

We claim that

1. $\sum_{i=1}^{k} c(x_i)$ is no more than the total charge on A^*.
2. The total charge on $e \in A^*$ is at most $H(\gamma) c(e)$.

The first claim is true because

$$\sum_{i=1}^{k} c(x_i) = \sum_{i=1}^{k} \mu_i \Delta_{x_i} f(A_{i-1})$$

$$= \sum_{i=1}^{k} \mu_i \left(f(A_i) - f(A_{i-1}) \right)$$

$$= \sum_{i=1}^{k} \mu_i \left((f(A^*) - f(A_{i-1})) - (f(A^*) - f(A_i)) \right)$$

$$= \sum_{i=1}^{k} (\mu_i - \mu_{i-1}) \left(f(A^*) - f(A_{i-1}) \right)$$

$$\leq \sum_{i=1}^{k} (\mu_i - \mu_{i-1}) \sum_{e \in S} \Delta_e f(A_{i-1})$$

$$= \sum_{e \in S} \sum_{i=1}^{k} (\mu_i - \mu_{i-1}) \Delta_e f(A_{i-1})$$

$$= \sum_{e \in S} \sum_{i=1}^{k} \mu_i \left(\Delta_e f(A_{i-1}) - \Delta_e f(A_i) \right).$$

Next, we prove the second claim. Consider an arbitrary element $e \in A^*$. Let l be the first i such that $\Delta_e f(A_i) = 0$. For each $1 \leq i \leq l$, by the greedy rule,

$$\mu_i = \frac{c(x_i)}{\Delta_{x_i} f(A_{i-1})} \leq \frac{c(e)}{\Delta_e f(A_{i-1})}.$$

Hence,

$$\sum_{i=1}^{k} \mu_i \left(\Delta_e f(A_{i-1}) - \Delta_e f(A_i) \right)$$

$$= \sum_{i=1}^{l-1} \mu_i \left(\Delta_e f(A_{i-1}) - \Delta_e f(A_i) \right) + \mu_l \Delta_e f(A_{l-1})$$

$$\leq c(e)\left(1+\sum_{i=1}^{l-1}\frac{\Delta_e f(A_{i-1}) - \Delta_e f(A_i)}{\Delta_e f(A_{i-1})}\right)$$

$$\leq c(e)\left(1+\sum_{i=1}^{l-1}(H(\Delta_e f(A_{i-1})) - H(\Delta_e f(A_i)))\right)$$

$$= c(e)(1 + H(\Delta_e f(\emptyset)) - H(\Delta_e f(A_{l-1})))$$

$$\leq c(e)(1 + H(\gamma) - H(1))$$

$$= c(e)H(\gamma).$$

So, the second claim also holds.

The two claims imply that

$$\sum_{i=1}^{k} c(x_i) \leq H(\gamma) \sum_{e \in S} c(e). \qquad \square$$

Now, we return to MIN-WCDS and note the following.

Lemma 2.4.4. $|V| - q(A)$ *is a submodular, monotone increasing integer function with* $|V| - q(\emptyset) = 0$.

Proof. Note that q is an integer function and $q(\emptyset)$. By Lemma 2.4.2, it suffices to show that for any $v \in V$,

$$A \subset B \Rightarrow \Delta_v q(A) \leq \Delta_v q(B).$$

Let $H(A)$ denote the graph with vertex set V and edge set consisting of all edges incident to a vertex in A. Then, each connected component of graph $H(B)$ is constituted by one or more connected components of graph $H(A)$. Thus, the number of connected components of $H(B)$ adjacent to v is no more than the number of connected components of $H(A)$ adjacent to v. Therefore, the lemma holds. $\qquad \square$

If we set $f(A) = |V| - q(A)$, then it is easy to see that MIN-SUBMODULAR-COVER becomes MIN-WCDS, Wolsey Greedy Algorithm becomes Chen–Liestman Algorithm, and Theorem 2.4.1 can result from Wolsey Theorem.

2.5 One-Stage Greedy Approximation

When a potential function is not submodular, how do we analyze a greedy algorithm with it? We study this problem in this section.

To avoid the above counterexample, we replace $h(C)$ by $q(C)$ the number of connected components of the subgraph with vertex set V and edge set $D(C)$, where $D(C)$ be the set of all edges incident to vertices in C. Define $f(C) = p(C) + q(C)$.

2.5 One-Stage Greedy Approximation

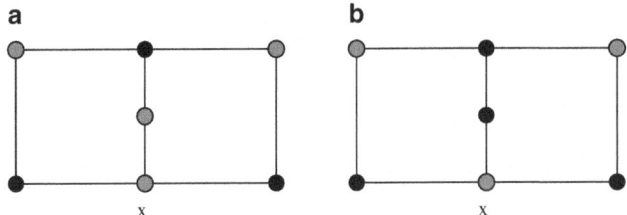

Fig. 2.3 A counterexample ($\Delta_x f(A) > \Delta_x f(B)$ but $A \subset B$)

Lemma 2.5.1. *Suppose G is a connected graph with at least three vertices. Then, C is a CDS if and only if $f(C \cup \{x\}) = f(C)$ for every $x \in V$.*

Proof. If C is a CDS, then $f(C) = 2$, which reaches the minimum value. Therefore, $f(C \cup \{x\}) = f(C)$ for every $x \in V$.

Conversely, suppose $f(C \cup \{x\}) = f(C)$ for every $x \in V$. First, C cannot be the empty set. In fact, for contradiction, suppose $C = \emptyset$. Since G is a connected graph with at least three vertices, there must exist a vertex x with degree at least two and for such a vertex x, $f(C \cup \{x\}) < f(C)$, a contradiction. Now, we may assume $C \neq \emptyset$. Consider a connected component of the subgraph induced by C. Let B denote its vertex set which is a subset of C. For every gray vertex y adjacent to B, if y is adjacent to a white vertex or a gray vertex not adjacent to B, then we must have $p(C \cup \{y\}) < p(C)$ and $q(C \cup \{y\}) \leq q(C)$; if y is adjacent to a black vertex not in B, then $p(C \cup \{y\}) \leq p(C)$ and $q(C \cup \{y\}) < q(C)$; hence, in all cases $f(C \cup \{y\}) < f(C)$, a contradiction. Therefore, every gray vertex adjacent to B cannot be adjacent to any vertex neither in B nor adjacent to B. Since G is connected, it follows that every vertex of G must belong to B or adjacent to B. That is, $B = C$ is a CDS. □

This lemma means that with $-f$ as potential function, Wolsey Greedy Algorithm would produce a CDS. If f is a monotone decreasing, submodular function, then we could directly employ Wolsey Theorem to give an estimation on performance ratio of the algorithm. Unfortunately, f is not submodular. A counterexample is shown in Fig. 2.3.

Could we also give analysis of Wolsey Greedy Algorithm in this case? The answer is yes and a new technique can be introduced based on two observations in the following.

The first observation is that MIN-CDS is an unweighted problem and in unweighted case, there is a simpler analysis for Wolsey Greedy Algorithm.

A Simple Analysis of Wolsey Greedy Algorithm in Unweighted Case: Let $x_1, \ldots, x_g$ be subsets selected in turn by Wolsey Greedy Algorithm. Denote $A_i = \{x_1, \ldots, x_i\}$. Let opt be the number of subsets in a minimum submodular cover. Let $C = \{y_1, \ldots, y_{\text{opt}}\}$ be a minimum submodular cover. Denote $C_j = \{y_1, \ldots, y_j\}$.

By the greedy rule,

$$f(A_{i+1}) - f(A_i) = \Delta_{x_{i+1}} f(A_i) \geq \Delta_{y_j} f(A_i)$$

for $1 \leq j \leq \text{opt}$. Therefore,
$$f(A_{i+1}) - f(A_i) \geq \frac{\sum_{j=1}^{\text{opt}} \Delta_{y_j} f(A_i)}{\text{opt}}.$$

On the other hand,
$$\frac{f(C) - f(A_i)}{\text{opt}} = \frac{f(A_i \cup C) - f(A_i)}{\text{opt}}$$
$$= \frac{\sum_{j=1}^{\text{opt}} \Delta_{y_j} f(A_i \cup C_{j-1})}{\text{opt}}.$$

Because f is submodular and monotone increasing, we have
$$\Delta_{y_j} f(A_i) \geq \Delta_{y_j} f(A_i \cup C_{j-1}).$$

Therefore,
$$f(A_{i+1}) - f(A_i) \geq \frac{f(C) - f(A_i)}{\text{opt}}, \tag{2.1}$$

that is,
$$f(C) - f(A_{i+1}) \leq (f(C) - f(A_i))\left(1 - \frac{1}{\text{opt}}\right)$$
$$\leq (f(C) - f(\emptyset))\left(1 - \frac{1}{\text{opt}}\right)^{i+1}$$
$$\leq (f(C) - f(\emptyset))e^{-(i+1)/\text{opt}}.$$

Choose i such that $f(C) - f(A_{i+1}) < \text{opt} \leq f(C) - f(A_i)$. Then
$$0 = f(C) - f(A_g) < f(C) - f(A_{g-1}) < \cdots < f(C) - f(A_{i+1}) \leq \text{opt} - 1.$$

Therefore,
$$g \leq i + \text{opt}$$
and
$$\text{opt} \leq (f(C) - f(\emptyset))e^{-i/\text{opt}}.$$

Therefore,
$$g \leq \text{opt}\left(1 + \ln \frac{f(C) - f(\emptyset)}{\text{opt}}\right) \leq \text{opt}(1 + \ln \gamma)$$

since
$$f(C) - f(\emptyset) \leq \sum_{j=1}^{\text{opt}} \Delta_{y_j} f(\emptyset) \leq \text{opt} \cdot \gamma. \qquad \square$$

2.5 One-Stage Greedy Approximation

The second observation is that in this analysis, there is only one place that submodularity is required, which is in the proof of inequality (2.1), where we need to have

$$\Delta_{y_j} f(A_i) \geq \Delta_{y_j} f(A_i \cup C_{j-1}).$$

An important observation on this inequality is that the increment variable y_j belongs to optimal solution. Therefore, although for nonsubmodular f this inequality may not holds, we may choose a proper ordering for things in optimal solution to make this inequality almost holds. In the following, we will implement this idea for CDS.

Let vertices $x_1, \ldots, x_g$ be selected in turn by Wolsey Greedy Algorithm. Denote $C_i = \{x_1, x_2, \ldots, x_i\}$ and $a_i = f(C_i)$. Initially, $a_0 = n$ where n is the number of vertices in G. Let C^* be a minimum CDS for G.

Lemma 2.5.2. *For* $i = 1, 2, \ldots, g$,

$$a_i \leq a_{i-1} - \frac{a_{i-1} - 2}{|C^*|} + 1.$$

Proof. First, consider $i \geq 2$. Note that

$$a_i = f(C_i) = a_{i-1} + \Delta_{x_i} f(C_{i-1}),$$

where

$$-\Delta_{x_i} f(C_{i-1}) = \max_y (-\Delta_y f(C_{i-1})).$$

Since C^* is a CDS, we can always arrange elements of C^* in an ordering $y_1, y_2, \ldots, y_{|C^*|}$ such that y_1 is adjacent to a vertex in C_{i-1} and for $j \geq 2$, y_j is adjacent to a vertex in $\{y_1, \ldots, y_{j-1}\}$. Denote $C_j^* = \{y_1, y_2, \ldots, y_j\}$. Then

$$\Delta_{C^*} f(C_{i-1}) = \sum_{j=1}^{|C^*|} \Delta_{y_j} f(C_{i-1} \cup C_{j-1}^*).$$

Note that

$$-\Delta_{y_j} p(C_{i-1} \cup C_{j-1}^*) \leq -\Delta_{y_j} p(C_{i-1}) + 1.$$

In fact, y_j can dominate at most one additional connected component in the subgraph $G[C_{i-1} \cup C_{j-1}^*]$ than in $G[C_{i-1}]$, which is the one contains C_{j-1}^* since $y_1, \ldots, y_{j-1}$ are connected. Moreover, by Lemma 2.4.4,

$$-\Delta_{y_j} q(C_{i-1} \cup C_{j-1}^*) \leq -\Delta_{y_j} q(C_{i-1}).$$

Therefore,

$$-\Delta_{y_j} f(C_{i-1} \cup C_{j-1}^*) \leq -\Delta_{y_j} f(C_{i-1}) + 1.$$

It follows that

$$a_{i-1} - 2 = -\Delta_{C^*} f(C_{i-1})$$
$$\leq \sum_{j=1}^{|C^*|} (-\Delta_{y_j} f(C_{i-1}) + 1).$$

There exists $y_j \in C^*$ such that

$$-\Delta_{y_j} f(C_{i-1}) + 1 \geq \frac{a_{i-1} - 2}{|C^*|}.$$

Hence,

$$-\Delta_{x_i} f(C_{i-1}) \geq \frac{a_{i-1} - 2}{|C^*|} - 1.$$

It implies that

$$a_i \leq a_{i-1} - \frac{a_{i-1} - 2}{|C^*|} + 1.$$

For $i = 1$, the proof is similar, we only need to note a difference that y_1 can be chosen arbitrarily. □

Theorem 2.5.3 (Ruan et al. [92]). *Wolsey Greedy Algorithm with $-f = -p - q$ as a potential function gives a polynomial-time $(2 + \ln \delta)$-approximation for* MIN-CDS *where δ is the maximum vertex-degree in input graph.*

Proof. By Lemma 2.5.2,

$$a_i - 2 \leq (a_{i-1} - 2)\left(1 - \frac{1}{|C^*|}\right) + 1$$
$$\leq (a_0 - 2)\left(1 - \frac{1}{|C^*|}\right)^i + \sum_{k=0}^{i-1}\left(1 - \frac{1}{|C^*|}\right)^k$$
$$= (a_0 - 2)\left(1 - \frac{1}{|C^*|}\right)^i + |C^*|\left(1 - \left(1 - \frac{1}{|C^*|}\right)^i\right)$$
$$= (a_0 - 2 - |C^*|)\left(1 - \frac{1}{|C^*|}\right)^i + |C^*|.$$

Since $a_i \leq a_{i-1} - 1$ and $a_g = 2$, we have $a_{|C|-2|C^*|} \geq 2|C^*| + 2$. If $g \leq 2|C^*|$, then we already done the proof. If $g > 2|C^*|$, then set $i = g - 2|C^*|$. Then

$$2|C^*| \leq (n - 2 - |C^*|)\left(1 - \frac{1}{|C^*|}\right)^i + |C^*|.$$

2.5 One-Stage Greedy Approximation

Since $(1 - 1/|C^*|)^i \leq e^{-i/|C^*|}$, we obtain

$$i \leq |C^*| \ln \frac{n - 2 - |C^*|}{|C^*|}.$$

Note that each vertex can dominate at most $\delta + 1$ vertices. Hence, $n/|C^*| \leq \delta + 1$. Therefore, $g = i + 2|C^*| \leq |C^*|(2 + \ln \delta)$. □

Now, let us consider Wolsey Greedy Algorithm for MIN-SET-COVER. If in each iteration we allow to choose two subsets instead of only one subset, could this greedy algorithm get a better performance ratio? The answer is not. This happens also to Wolsey Greedy Algorithm for submodular potential function. However, for nonsubmodular potential function, the situation is changed. The following greedy algorithm will approach to performance ratio $1 + \ln \delta$ as $k \to \infty$.

Greedy Algorithm DGPWWZ
input a connected graph G.

    Initially, set $C \leftarrow \emptyset$;
    while $f(C) > 2$ **do**
        choose a subset X of at most $2k - 1$ vertices to maximize $-\frac{\Delta_X f(C)}{|X|}$
        and set $C \leftarrow C \cup X$;
output $C_g = C$.

To analyze Greedy Algorithm DGPWWZ, we need to note the following property of the potential function $-f$.

Lemma 2.5.4. *Let A and B be two vertex subsets. If both $G[B]$ and $G[X]$ are connected, then*

$$-\Delta_X f(A \cup B) + \Delta_X f(A) \leq 1.$$

Proof. Since q is submodular, we have $\Delta_X q(A) \leq \Delta_X q(A \cup B)$.

Moreover, since both subgraphs $G[B]$ and $G[X]$ are connected, the number of black components dominated by X in $G[A \cup B]$ is at most one more than the number of black components dominated by X in $G[A]$. Therefore, $-\Delta_X p(A \cup B) \leq -\Delta_X p(A) + 1$. Hence, $-\Delta_X f(A \cup B) \leq -\Delta_X f(A) + 1$. □

Let C^* be a minimum CDS. We show two properties of C^* in the following two lemmas.

Lemma 2.5.5. *For any integer $k \geq 2$, C^* can be decomposed into $Y_1, Y_2, \ldots, Y_h$ for some natural number h such that*

(a) $C^* = Y_1 \cup Y_2 \cup \cdots \cup Y_h$.
(b) *For $1 \leq i \leq h$, both $G[Y_1 \cup Y_2 \cup \cdots \cup Y_i]$ and $G[Y_i]$ are connected.*
(c) $k + 1 \leq |Y_i| \leq 2k - 1$ *for $1 \leq i \leq h$ except one, in such an exceptional i, $1 \leq |Y_i| \leq 2k - 1$.*
(d) $|Y_1| + |Y_2| + \cdots + |Y_h| \leq |C^*| + h - 1$.

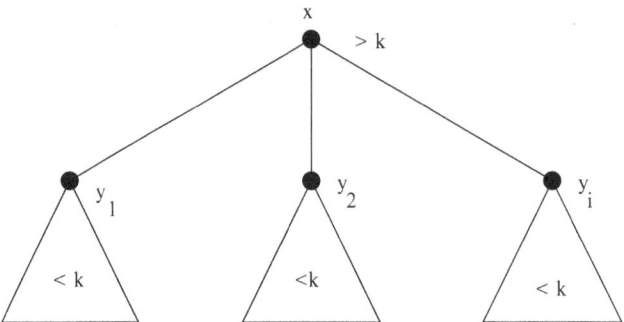

Fig. 2.4 Case 2 in proof of Lemma 2.5.5

Proof. Consider a tree T with vertex set C^*. Choose a vertex $r \in C^*$ as the root of T. For any vertex $x \in C^*$, let $T(x)$ denote the subtree rooted at x and $|T(x)|$ the number of vertices in $T(x)$. If T contains more than $2k-1$ vertices, then there must exist a vertex $x \in C^*$ such that $|T(x)| \geq k+1$ and for every child y of x, $|T(y)| \leq k$. Next, consider two cases.

Case 1. There is a child y of x such that $|T(y)| = k$. Let Y_1 consist of all vertices of $T(y)$ together with x and delete all vertices of $T(y)$ from T.

Case 2. For every child y of x, $|T(y)| \leq k-1$. Suppose $y_1, \ldots, y_i$ are all children of x (Fig. 2.4). There must exist $2 \leq j \leq i$ such that $|T(y_1)| + \cdots + |T(y_j)| \leq k-1$ and $|T(y_1)| + \cdots + |T(y_j)| + |T(y_{j+1})| \geq k$. Since $|T(y_{j+1})| \leq k-1$, we have $|T(y_1)| + \cdots + |T(y_j)| + |T(y_{j+1})| \leq 2k-2$. Let Y_1 consist all vertices in $T(y_1) \cup \cdots T(y_{j+1})$ together with x and delete $Y_1 - \{x\}$ from T.

Repeating above process on the remainder of T, we will obtain a required decomposition. □

Lemma 2.5.6. *Let $n = |V|$. Then, $n \leq (\delta - 1)|C^*| + 2$.*

Proof. We prove by induction on $|C^*|$ that C^* with connected $G[C^*]$ can dominate at most $(\delta - 1)|C^*| + 2$ vertices. For $|C^*| = 1$, it is trivially true. For $|C^*| \geq 2$, choose a vertex $x \in C^*$ such that $G[C^* - \{x\}]$ is still connected. Removal x would remove at most $\delta - 1$ vertices from the set of vertices dominated by C^*. By the induction hypothesis, $C^* - \{x\}$ can dominate at most $(\delta - 1)(|C^*| - 1) + 2$ vertices. Therefore, C^* can dominate at most $(\delta - 1)|C^*| + 2$ vertices. □

Theorem 2.5.7 (Du et al. [40]). *For any $\varepsilon > 0$, there exists a polynomial-time approximation with performance ratio $(1 + \varepsilon)\ln(\delta - 1)$ for MIN-CDS.*

Proof. Note that the input graph G is connected. If its maximum degree $\delta = 1$, then G contains only one edge. This means that C_g contains only one vertex and is optimal. Hence, Theorem 2.5.7 holds. For $\delta = 2$, G is a path or a cycle. When G is a path, its minimum CDS consists of all internal vertices. When G is a cycle, a minimum CDS can be obtained by deleting two adjacent vertices.

2.5 One-Stage Greedy Approximation

Hence, Theorem 2.5.7 holds. Therefore, we may assume $\delta \geq 3$. Under this assumption, we may further assume $|C_g| > 2|C^*|$ where C^* is a minimum CDS, since, otherwise, $|C_g| \leq 2|C^*| \leq (1+\ln 3)|C^*| \leq (1+c+\ln\delta)|C^*|$.

Suppose $X_1, \ldots, X_g$ are chosen by Greedy Algorithm 2.5 and denote $C_i = X_1 \cup \cdots \cup X_i$. Decompose a minimum CDS C_j^* into $Y_1, \ldots, Y_h$ satisfying conditions in Lemma 2.5.5. Denote $C_j^* = Y_1 \cup \cdots \cup Y_j$. By Lemmas 2.5.4 and 2.5.5,

$$-\Delta_{Y_j} f(C_i \cup C_{j-1}^*) = -\Delta_{Y_j} p(C_i \cup C_{j-1}^*) - \Delta_{Y_j} q(C_i \cup C_{j-1}^*)$$
$$\leq -\Delta_{Y_j} p(C_i) + 1 - \Delta_{Y_j} q(C_i)$$
$$\leq -\Delta_{Y_j} f(C_i) + 1.$$

By greedy rule,

$$\frac{-\Delta_{X_{i+1}} f(C_i)}{|X_{i+1}|} \geq \frac{-\Delta_{Y_j} f(C_i)}{|Y_j|} \text{ for } 1 \leq j \leq h.$$

Hence,

$$\frac{-\Delta_{X_{i+1}} f(C_i)}{|X_{i+1}|} \geq \frac{-\sum_{j=1}^h \Delta_{Y_j} f(C_i)}{\sum_{j=1}^h |Y_j|}$$
$$\geq \frac{-(h-1) - \sum_{j=1}^h \Delta_{Y_j} f(C_i \cup C_{j-1}^*)}{\sum_{j=1}^h |Y_j|}$$
$$\geq \frac{-(h-1) - (f(C_i \cup C^*) - f(C_i))}{\text{opt} + h - 1}$$
$$= \frac{f(C_i) - (h+1)}{\text{opt} + h - 1}$$

where $\text{opt} = |C^*|$. Denote $a_i = f(C_i - (h+1))$. Then,

$$\frac{a_i - a_{i+1}}{|X_{i+1}|} \geq \frac{a_i}{\text{opt} + h - 1},$$

that is,

$$a_{i+1} \leq a_i \left(1 - \frac{|X_{i+1}|}{\text{opt} + h_1}\right) \leq a_i e^{-|X_{i+1}|/(\text{opt}+h-1)}$$
$$\leq a_0 e^{-(|X_{i+1}|+|X_i|+\cdots+|X_1|)/(\text{opt}+h-1)}.$$

Choose i such that

$$a_{i+1} < \text{opt} \leq a_i.$$

Denote $b = a_i - \text{opt}$ and $b' = \text{opt} - a_{i+1}$. Write $|X_{i+1}| = d + d'$ such that

$$\frac{b}{d} = \frac{b'}{d'} = \frac{a_i - a_{i+1}}{|X_{i+1}|} \geq \frac{a_i}{\text{opt} + h - 1}.$$

Then we have

$$\frac{a_i - \text{opt}}{d} = \frac{b}{d} \geq \frac{a_i}{\text{opt} + h - 1}.$$

So,

$$\text{opt} \leq a_i \left(1 - \frac{d}{\text{opt} + h - 1}\right) \leq a_i e^{-d/(\text{opt}+h-1)}.$$

Hence

$$\text{opt} \leq a_0 e^{-(d + |X_i| + \cdots + |X_1|)/(\text{opt}+h-1)},$$

Note $a_0 = f(\emptyset) - (h+1) = n - (h+1)$. Thus,

$$|X_1| + \cdots + |X_i| + d \leq (\text{opt} + h - 1) \ln \frac{n - (h+1)}{\text{opt}}.$$

Moreover,

$$d' + |X_{i+2}| + \cdots + |X_g| \leq b' + f(C_{i+1}) - f(C_g)$$
$$= \text{opt} - a_{i+1} + f(C_{i+1}) - f(C^*)$$
$$= \text{opt} + (h - 1).$$

Therefore,

$$|X_1| + \cdots + |X_g| \leq \text{opt}\left(1 + \frac{1}{k}\left(1 + \ln \frac{n - (h+1)}{\text{opt}}\right)\right).$$

By Lemma 2.5.6, $n \leq (\delta - 1)\text{opt} + 2$. Since $h \geq 1$, we have

$$\frac{n - (h+1)}{\text{opt}} \leq \delta - 1.$$

Hence,

$$|X_1| + \cdots + |X_g| \leq \left(1 + \frac{1}{k}\right)(1 + \ln(\delta - 1)).$$

Choose k such that $1/k < \varepsilon$. We obtain Theorem 2.5.7. □

2.6 Weighted CDS

Consider a graph $G = (V, E)$ with weight $w : V \to R^+$. In Chap. 1, we discussed the relationship between CDS and leaves of a spanning tree. From the discussion, we can easily see that a vertex subset is a CDS if and only if it contains all internal vertices of a spanning tree. Therefore, we can obtain the following facts:

- G has a CDS with minimum weight w^* if and only if G has a spanning tree with minimum total internal vertex weight w^*.
- G has a CDS with weight at most w if and only if G has a spanning tree with total internal vertex weight at most w.

Given a digraph and a source vertex s, a *broadcasting tree* is a tree with root s and containing paths that from s to each vertex in the digraph. The broadcasting tree is also called an *out-arborescence*. When we treat G as a digraph by replacing each edge with two arcs with different directions, the following relationship between broadcasting tree and CDS follows from above relationship between spanning tree and CDS:

- G has a CDS with minimum weight w^* if and only if there exists a source vertex s such that G has a broadcasting tree from s with minimum total internal vertex weight w^*.
- G has a CDS with weight at most w if and only if there exists a source vertex s such that G has a broadcasting tree with total internal vertex weight at most w.

Due to this relationship, we first study the following problem on broadcasting tree.

MINW-BROADCAST: Given a digraph $G = (V, E)$ with weight $w : V \to R^+$ and a source node, find a broadcasting tree with minimum total weight of internal nodes.

Consider a subgraph H of input graph G with a source s. An *orphan* of H is a strongly connected component without coming edge and not containing s. Given each node an individual integer ID, the node with smallest ID in an orphan is called the *head* of the orphan.

A *spider* is a subgraph consisting of a body node and several directed paths from the body node to its feet (see Fig. 2.5). A spider is *legal* if it satisfies three conditions:

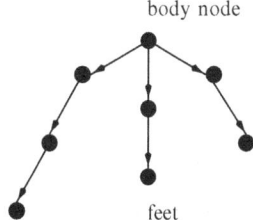

Fig. 2.5 Spider

Fig. 2.6 A new orphan is produced by adding a spider

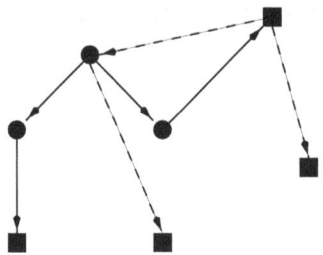

1. All feet are heads of some orphans.
2. S head in it must be a foot or body node.
3. Either its body node is the source s or it contains at least two orphan heads.

We ask for the second condition because putting a legal spider in H may introduce a new orphan at the body node when the body node is not source s so that the number of reduced orphan heads should be the number of orphan heads in it minus one (Fig. 2.6).

For a legal spider S, let $h(S)$ be the number of orphan heads in S and $\mathrm{cost}(S)$ the total weight of internal nodes in S other than internal nodes in H. Define

$$\mathrm{quotient}(S) = \frac{\mathrm{cost}(S)}{h(S)}.$$

For any node u, cut at all orphan heads and consider the connected component C containing u. Suppose $p_1, \ldots, p_k$ are k shortest paths from u to k different orphan heads in C. We consider $S = p_1 \cup \cdots \cup p_k$ as a spider although $p_1, \ldots, p_k$ may have some common nodes other than u. When calculate $\mathrm{cost}(S)$, we assume that all $p_1, \ldots, p_k$ are disjoint except at body node. Therefore, $\mathrm{cost}(S)$ is actually an upper bound for the total weight of increased internal nodes. The purpose to make this assumption is to have an easy way to compute $\mathrm{quotient}(v)$ for every node u, which is defined to be

$$\mathrm{quotient}(u) = \min\{\mathrm{quotient}(S) \mid S \text{ is over all legal spider with body node } u\}.$$

With above assumption, $\mathrm{quotient}(u)$ for any node u can be computed in the following way: Suppose H has k orphan heads and $p_1, \ldots, p_k$ are shortest paths from node u to them, respectively, ordering that $\mathrm{cost}(p_1) \leq \mathrm{cost}(p_2) \leq \cdots \leq \mathrm{cost}(p_k)$. Then for $u \neq s$,

$$\mathrm{quotient}(u) = \min_{2 \leq i \leq k} \mathrm{quotient}(p_1 \cup \cdots \cup p_i),$$

and for $u = s$,

$$\mathrm{quotient}(u) = \min_{1 \leq i \leq k} \mathrm{quotient}(p_1 \cup \cdots \cup p_i).$$

Before state the algorithm, let us show a useful lemma about $\mathrm{quotient}(u)$.

2.6 Weighted CDS

Fig. 2.7 Spider decomposition

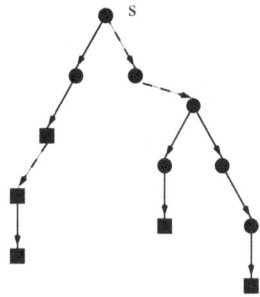

Lemma 2.6.1. *Let q be the number of orphans in H. Then there exists a node u with*

$$\text{quotient}(u) \leq \frac{\text{opt}}{q},$$

where opt *is the objective function value of optimal solution.*

Proof. Let T^* be an optimal broadcasting tree. We can prune T^* to obtain a subtree T such that every leaf is an orphan head. Now, we can obtain a sequence of legal spiders, $S_1, \ldots, S_\ell$ from decomposition of T (Fig. 2.7). Those legal spiders contains all orphan heads and all internal nodes either in H or in T. Therefore,

$$\text{cost}(S_1) + \cdots + \text{cost}(S_\ell) \leq \text{opt}$$

and

$$h(S_1) + \cdots + h(S_\ell) = q.$$

Thus,

$$\min_{1 \leq i \leq \ell} \text{quotient}(S_i) \leq \frac{\text{opt}}{q}.$$

This means that one of heads for $S_1, \ldots, S_\ell$ meet our requirement. □

Algorithm Broadcast:
input a strongly connected digraph $G = (V, E)$ with source node s;
$U \leftarrow \{s\}$;
$O \leftarrow V - \{s\}$;
while $O \neq \emptyset$ **do begin**
      choose node u with smallest quotient cost;
      let $S(u)$ be the legal spider at u reaching quotient(u);
      $U \leftarrow U \cup S(u)$;
      remove from O those orphans whose heads in $S(u)$
      and add back possibly one new orphan;
      recalculate quotient cost of each node;
end-while
output U.

Theorem 2.6.2 (Li et al. [76]). MINW-BROADCAST *has a polynomial-time* $(1+2\ln(n-1))$-*approximation.*

Proof. We analyze the Algorithm Broadcast. Suppose the algorithm runs in k iterations. Initially, there are $n_0 = n-1$ orphans. Let n_i denote the number of orphans right after the ith iteration. For $1 \leq i \leq k$, let S_i be the legal spider chosen at the ith iteration. Let h_i be the number of heads in S_i and $c_i = \mathrm{cost}(S_i)$. Then

$$n_i \leq n_{i-1} - \frac{h_i}{2},$$

since if $h_i = 1$, then

$$n_i \leq n_{i-1} - 1 \leq n_{i-1} - \frac{h_i}{2};$$

and if $h_i \geq 2$, then

$$n_i \leq n_{i-1} - h_i + 1 \leq n_{i-1} - \frac{h_i}{2}.$$

Moreover, by Lemma 2.6.1,

$$\frac{c_i}{h_i} \leq \frac{\mathrm{opt}}{n_{i-1}}.$$

Thus,

$$\frac{n_i}{n_{i-1}} \leq 1 - \frac{c_i}{2\mathrm{opt}}.$$

It implies that

$$\frac{n_{k-1}}{n_0} \leq \prod_{i=1}^{k-1}\left(1 - \frac{c_i}{2\mathrm{opt}}\right).$$

Hence,

$$\ln \frac{n_{k-1}}{n_0} \leq -\frac{c_1 + \cdots c_{k-1}}{2\mathrm{opt}},$$

that is,

$$c_1 + \cdots + c_{k-1} \leq 2\mathrm{opt} \cdot \ln\frac{n_0}{n_{k-1}} \leq 2\mathrm{opt} \cdot \ln(n-1).$$

Since $\frac{c_k}{h_k} \leq \frac{\mathrm{opt}}{n_{k-1}}$ and $h_k = n_{k-1}$, we have $c_k \leq \mathrm{opt}$. Therefore,

$$c_1 + \cdots + c_k \leq (1 + 2\ln(n-1)) \cdot \mathrm{opt}. \qquad \square$$

Now, we return to MINW-CDS.

Theorem 2.6.3. MINW-CDS *has a polynomial-time* $(1+2\ln(n-1))$-*approximation where n is the number of nodes in input graph.*

Proof. Suppose $G = (V, E)$ is an input graph with weight $w : V \to R^+$. Choose a node $u \in V$. Let $N(u)$ denote the set of neighbors of u and u. For each $v \in N(u)$, compute a broadcasting tree T_v with source v by Algorithm Broadcast. From those T_v for $v \in N(u)$, choose T_{v^*} with minimum total weight of internal nodes. We show that all internal nodes of T_{v^*} form a CDS C with total weight within a factor of $1 + 2\ln(n-1)$ from optimal.

Let C^* be a CDS with minimum total weight w^*. Note that $C^* \cap N(u) \neq \emptyset$. Choose $v \in C^* \cap N(u)$. Construct a spanning tree for $G[C^*]$ and extend it to a spanning tree for G. Give each edge a direction to form a broadcasting tree T_v^* from source v. Then, T_v^* has total internal node weight at most w^*. By Theorem 2.6.2,

$$\text{weight}(T_v^*) \leq (1 + 2\ln(n-1))\text{weight}(T_v^*),$$

where $\text{weight}(T_v)$ denotes the total internal node weight of T_v. Therefore,

$$\text{weight}(C) \leq \text{weight}(T_v^*) \leq (1 + 2\ln(n-1))\text{weight}(T_v^*) \leq (1 + 2\ln(n-1))w^*. \quad \Box$$

2.7 Directed CDS

In this section, we show a relationship between SCDS and the broadcast tree.

Lemma 2.7.1. *Let $\text{opt}_{\text{BT}}(G, r)$ be the objective function value of optimal solution for* MINW-BROADCAST *on input G and a source r. Let $\text{opt}_{\text{SCDS}}(G)$ be the objective function value of an optimal solution for* MIN-SCDS *on input G. Then for any r,*

$$\text{opt}_{\text{BT}}(G, r) \leq \text{opt}_{\text{SCDS}}(G).$$

Moreover, if r belongs to an optimal solution for MIN-SCDS, *then*

$$\text{opt}_{\text{BT}}(G, r) \leq \text{opt}_{\text{SCDS}}(G) - 1.$$

Proof. Let C^* be the minimum SCDS of G. For any resource r, we can first get in C^* and then through C^* to reach other nodes not in C^* so that the broadcasting tree uses only nodes in C^* as internal nodes except r. When $r \in C^*$, r can be taken off in counting $\text{opt}_{\text{BT}}(G, r)$. $\quad \Box$

Lemma 2.7.2. *If there exists polynomial-time α-approximation for* MINW-BROADCAST, *then there exists polynomial-time 2α-approximation for* MIN-SCDS.

Proof. Let G^R be a directed graph obtained from G by reversing the direction of each edge. Let C^* be a minimum strongly SCDS. Choose a node u arbitrarily. Let $N(u)$ be the set consisting of the node u and its in-neighbors, that is those nodes each of which has an edge coming to u. Clearly, $N(u) \cap C^* \neq \emptyset$. For each $s \in N(u)$, compute an α-approximation T_1 for MINW-BROADCAST on input G and source s and also a α-approximation T_2^R for MINW-BROADCAST on input G^R with source s. Let T_2 be the tree obtained from T_2^R by reversing the direction of each edge. Then $T_1 \cup T_2$ is a strongly connected spanning subgraph of G. Furthermore, $I(T_1) \cup I(T_2^R)$ induced a strongly connected subgraph of $T_1 \cup T_2$, dominating G. Hence $I(T_1) \cup I(T_2^R)$ is a SCDS for G where $I(T_i)$ denotes the set of internal nodes in T_i. Clearly,

$$|I(T_1) \cup I(T_2)| \leq |I(T_1) - \{s\}| + |I(T_2) - \{s\}| + |\{s\}|$$
$$\leq \alpha(\text{opt}_{\text{BT}}(G,s) + \text{opt}_{\text{BT}}(G^R,s)) + 1.$$

Note that when s belongs to a minimum SCDS, we would have

$$\alpha(\text{opt}_{\text{BT}}(G,s) + \text{opt}_{\text{BT}}(G^R,s)) + 1$$
$$\leq \alpha(\text{opt}_{\text{SCDS}}(G) - 1 + \text{opt}_{\text{SCDS}}(G^R) - 1) + 1$$
$$\leq 2\alpha \cdot \text{opt}_{\text{SCDS}}(G),$$

since a minimum SCDS for G is also a minimum SCDS for G^R, vice versa. Now, for s over all nodes in $N(u)$, we choose the one such that $I(T_1) \cup I(T_2^R)$ has the smallest cardinality. Such a $I(T_1) \cup I(T_2^R)$ will have cardinality upper bounded by $2\alpha \cdot \text{opt}_{\text{SCDS}}(G)$. □

By Lemma 2.7.2 and Theorem 2.6.2, we have

Theorem 2.7.3 (Li et al. [76]). *There exists a polynomial-time $(2 + 4\ln(n-1))$-approximation for* MIN-SCDS.

Chapter 3
CDS in Unit Disk Graph

> *Every dance is kind of fever chart, a graph of the heart.*
> MARTHA GRAHAM

3.1 Motivation and Overview

A unit disk is a disk with diameter one. Denote by $\text{disk}_r(o)$ the disk with center o and radius r. A graph $G = (V, E)$ is called a *unit disk graph* if it can be embedded into the Euclidean plane such that an edge between two nodes u and v exists if and only if $\text{disk}_{0.5}(u) \cap \text{disk}_{0.5}(v) \neq \emptyset$, that is, their Euclidean distance $d(u,v) \leq 1$. The unit disk graph is a mathematical model for wireless sensor networks when all sensors have the same communication radius.

For any node v of a unit disk graph G, the *neighborhood area* of v is the disk $\text{disk}_1(v)$. For any subset V' of nodes, the *neighborhood area* of V' is the union of disks, $\cup_{v \in V'} \text{disk}_1(v)$. For any subgraph H, the *neighborhood area* of H is the union of disks, $\cup_{v \in V(H)} \text{disk}_1(v)$ where $V(H)$ is the node set of subgraph H. Clearly, in a unit disk graph, two nodes u and v are independent if and only if $d(u,v) > 1$. For any two points u and v in the Euclidean plane, if $d(u,v) > 1$, then u and v are also said to be independent.

The boundary of an area Ω is denoted by $\partial \Omega$. Thus, $\partial \text{disk}_r(v) = \text{circle}_r(v)$, which is the circle with radius r and center v.

Clark, Colbourn, and Johnson [24] proved that MIN-CDS in unit disk graphs is still NP-hard. Wan et al. [104] first found that MIN-CDS has polynomial-time constant-approximations. Cheng et al. [22] designed the first PTAS. Since the running time of PTAS is a polynomial of a high degree, which is hard to implement, the design of fast polynomial-time constant-approximation is still an active research topic in the literature [14, 21, 51, 72, 75, 79, 104, 106]. There are many designs using the approach initiated by Wan et al. [104]: At the first stage, construct a maximal independent set. At the second stage, connect the maximal independent set into a CDS.

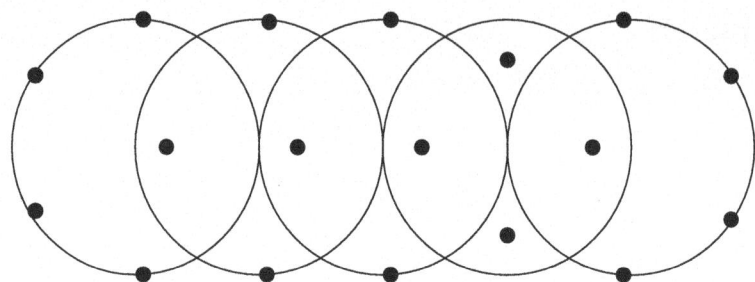

Fig. 3.1 The neighborhood of $n(\geq 3)$ linear points with consecutive distance one may contain $3n+3$ independent points

To analyze such two-stage algorithms, one needs to know what is the maximum size of a maximal independent set (i.e., the size of the maximum independent set) compared with the size of the minimum CDS. The size of the maximum independent set in a graph G is called the independent number, denoted by $\alpha(G)$. The size of the minimum CDS in G is called the connected dominating number, denoted by $\gamma_c(G)$. Wan et al. [106] indicated that there exist some connected unit disk graphs G such that

$$\alpha(G) = 3 \cdot \gamma_c(G) + 3$$

(Fig. 3.1). Many researchers believe that for every connected unit disk graph G

$$\alpha(G) \leq 3 \cdot \gamma_c(G) + 3. \tag{3.1}$$

Many efforts have been made to attack this upper bound. They can be classified into classes based on the difference of basic approaches.

One is based on the study of packing independent points in the neighborhood area of a small subgraph. Wan et al. [104] showed that the neighborhood area of any node can contain at most five independent points (Fig. 3.2) and based on this fact, they showed that for every connected unit disk graph G,

$$\alpha(G) \leq 4 \cdot \gamma_c(G) + 1.$$

Wu et al. [123] showed that the neighborhood area of any edge can contain at most eight independent points (Fig. 3.3), and with this fact, they showed that for every connected unit disk graph G

$$\alpha(G) \leq 3.8 \cdot \gamma_c(G) + 1.2.$$

Along this direction, Wan et al. [106] studied the neighborhood area of a star and proved that for every connected unit disk graphs with at least two nodes

$$\alpha(G) \leq 3\frac{2}{3} \cdot \gamma_c(G) + 1.$$

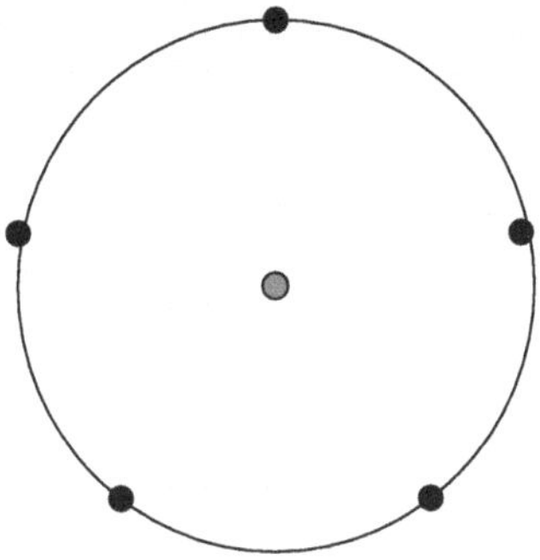

Fig. 3.2 A disk with radius one can contain at most five independent points

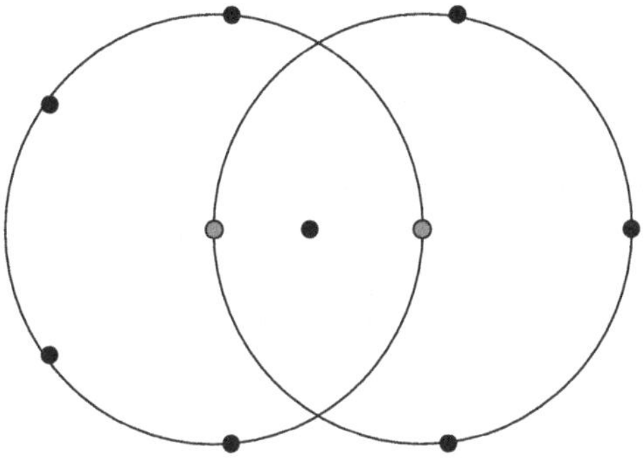

Fig. 3.3 The union of two disks $disk_1(u) \cap disk_1(v)$ with $d(u,v) \leq 1$ can contain at most eight independent points

Vahdatpour et al. [103] claimed that they proved (3.1). However, their proof is far from a complete one. An analysis on their proof will be given in Sect. 3.4.

Another approach is to study the total area taken by unattached unit disks in the union of disks of radius 1.5 and with centers at nodes in a CDS. With this approach and Voronoi division, Funke et al. [51] showed that for every connected unit disk graph G

$$\alpha(G) \leq 3.453 \cdot \gamma_c(G) + 8.291.$$

However, in their proof, a key geometric extreme property was used without proof. Therefore, some researchers could not accept this result. Gao et al. [54] gave a detail proof of the geometric extreme property. Li et al. [72] improved approach of Funke et al. and showed that for every connected unit disk graph G,

$$\alpha(G) \leq 3.4306 \cdot \gamma_c(G) + 4.8185.$$

This is the best-known bound so far.

3.2 NP-Hardness and PTAS

In this section, we give a new proof of NP-hardness and a new construction of PTAS for MIN-CDS in unit disk graphs.

Fig. 3.4 (a) A planar graph. (b) The constructed graph. The *dark circled* points are candidates of Steiner points and the *light circled* points are terminals

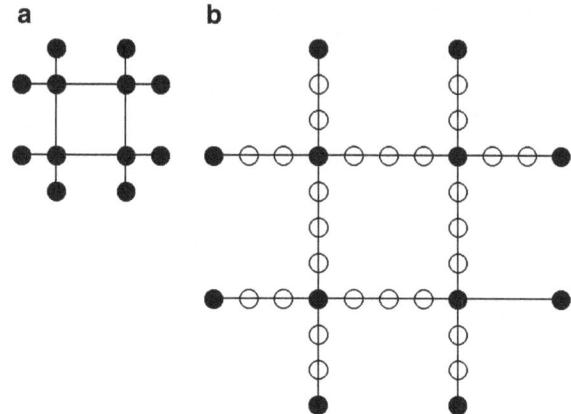

Theorem 3.2.1 (Clark et al. [24]). MIN-CDS *in unit disk graphs is still NP-hard.*

Proof. The following NP-complete problem can be found in [56, 57].

> PLANAR-4-CVC: Given a planar graph $G = (V, E)$ with all vertices of degree at most 4, and a positive integer $k \leq |V|$, determine whether there is a *connected vertex cover* of size k, that is, a subset $V' \subseteq V$ with $|V'| = k$ such that for each edge $\{u, v\} \in E$ at least one of u and v belongs to V' and the subgraph induced by V' is connected.

Consider a graph $G = (V, E)$ and a positive integer k, which is an instance of this problem. We construct a unit disk graph as follows. First, note that we can embed G into the plane so that all edges consist of horizontal and vertical segments of lengths being an integer at least 4, so that every two edges meet at an angle of $90°$ or $180°$. Add new vertices on the interior of each edge in G to divide the edge into a path of many edges, each of length exactly one. Denote by W the set of all such new vertices. (See Fig. 3.4. New vertices are light circled points.)

Now, consider a horizontal path $(u, w_1, \ldots, w_h, v)$ obtained from an edge (u, v). Choose a constant $0 < c < 0.5(\sqrt{2} - 1)$. For each new vertex w_i, add another new vertex w_i' such that $d(w_i, w_i') = c$ and w_i' is above w_i if i is odd and below w_i if i is even (Fig. 3.5). This placement of w_i' implies that w_i' can connect to only w_i. Similarly, we can deal with path obtained from vertical edges. Denote by G' the constructed graph. Then every CDS C of G' must contain w_i. In fact, in order to dominate w_i', C must contain either w_i or w_i'. If C contains w_i', then w_i' has to connect other vertices in C through w_i. Therefore, we must have w_i in C.

Now, it is easy to see the following facts:

1. W is a dominating set of G'.
2. C is a connected vertex-cover of G if and only if $C \cup W$ is a CDS of G'.

Therefore, G has a connected vertex-cover of size at most k if and only if G' has a CDS of size at most $|W| + k$. □

Next, we give a new construction of a PTAS for MIN-CDS in unit disk graphs.

3.2 NP-Hardness and PTAS

Fig. 3.5 Add w_i''s

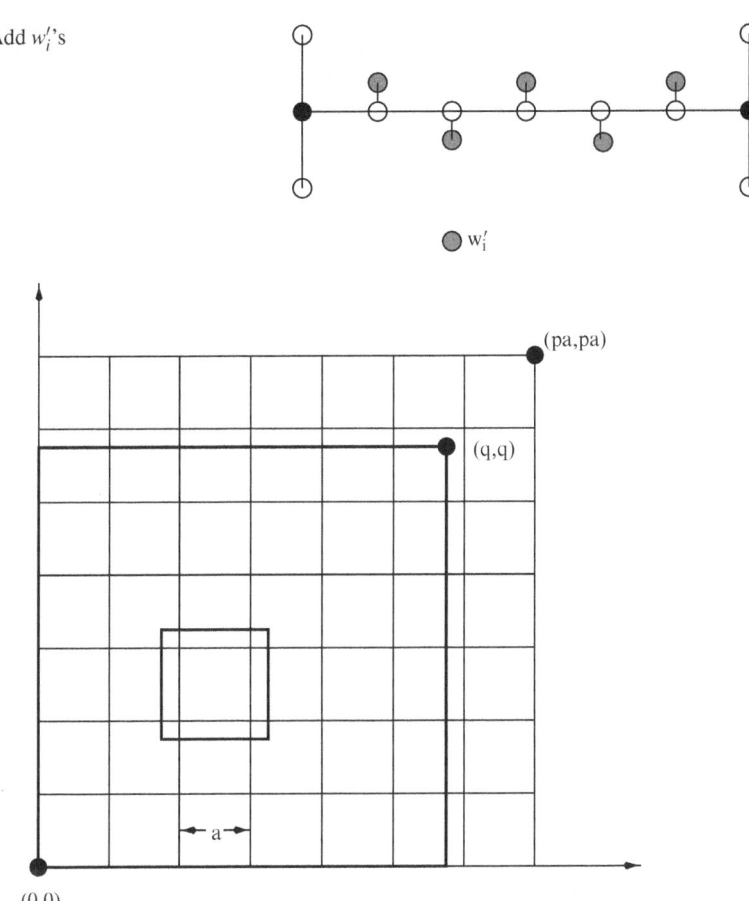

Fig. 3.6 Grid $P(0)$

Initially, we put input connected unit disk graph $G = (V, E)$ in the interior of the square $[0,q] \times [0,q]$ and construct a grid $P(0)$ as shown in Fig. 3.6. $P(0)$ divides the square $[0, pa] \times [0, pa]$ into p^2 cells where $a = 8k$ for a positive integer k and $p = 1 + \lceil q/a \rceil$. Each cell e is a $a \times a$ square, including its left boundary and its lower boundary, so that all cells are disjoint and their union covers the interior of the square $[0,q] \times [0,q]$.

For each cell e, let $C(e)$ be the closed area bounded by the $(a+4) \times (a+4)$ square with the same center as that of e, called the *central area* of cell e. Let $CB(e)$ be the interior of the $(a+8) \times (a+8)$ square with the same center as that of e. Denote by $B(e)$ obtained from $CB(e)$ by removing e, called the *boundary area* of cell e (Fig. 3.7).

Fig. 3.7 Central area $C(e)$ and Boundary area $B(e)$

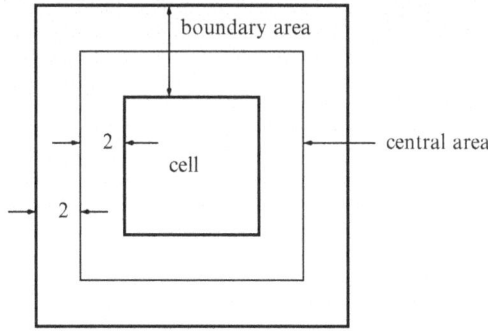

Fig. 3.8 Partition of central area $C(e)$

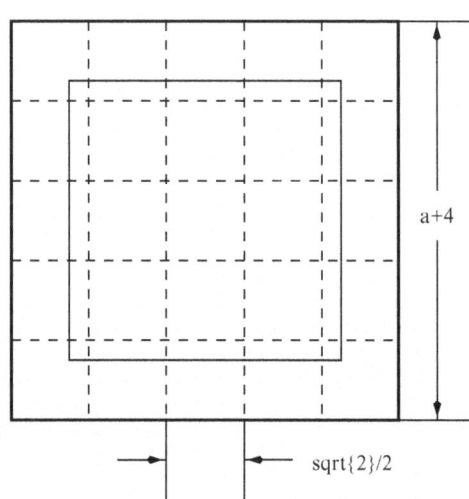

For each cell e, we study the following problem.

LOCAL(e): Find the minimum subset D of vertices in $V \cap CB(e)$ such that (a) D dominates all nodes in $V \cap C(e)$, and (b) for any connected component F of subgraph $G[V \cap C(e)]$, $G[D]$ contains a connected component dominating F.

Lemma 3.2.2. *The minimum solution of* LOCAL(e) *problem can be computed in time* $n_e^{O(a^2)}$ *where* $n_e = |V \cap CB(e)|$.

Proof. Partition $C(e)$ into $\lceil (a+4)\sqrt{2} \rceil^2$ small squares with edge length at most $\sqrt{2}/2$ (Fig. 3.8). Then for each closed small square s, if $V \cap s \neq \emptyset$, then choose one from $V \cap s$, which would dominate all vertices in $V \cap s$. All chosen vertices form a set D dominating $V \cap C(e)$ and $|D| \leq \lceil (a+4)\sqrt{2} \rceil^2$.

Now, consider each connected component F of $G[V \cap C(e)]$. If D does not contain a connected component dominating F, then we may add at most $2|F \cap D|$ vertices to connect all vertices $F \cap D$ into one connected component. This means that LOCAL(e) has a feasible solution of size at most $3\lceil (a+4)\sqrt{2} \rceil^2$. Therefore, we can find the optimal solution for LOCAL(e) in time $n_e^{3\lceil (a+4)\sqrt{2}\rceil^2} = n_e^{O(a^2)}$. □

3.2 NP-Hardness and PTAS

Let D_e denote the minimum solution for LOCAL(e). Define $D(0) = \cup_{e \in P(0)} D_e$ where $e \in P(0)$ means that e is over all cells in partition $P(0)$.

Lemma 3.2.3. *$D(0)$ contains a CDS for G, which can be computed in time $n^{O(a^2)}$.*

Proof. Consider two adjacent cells e and e'. Let F be a connected component of $G[V \cap C(e)]$ dominated by a connected component D_e^F of $G[D \cap CB(e)]$. Let F' be a connected component of $G[V \cap C(e')]$ dominated by a connected component $D_{e'}^{F'}$ of $G[D \cap CB(e')]$. Suppose $F \cup F'$ is connected. We claim that $G[D_e^F \cup D_{e'}^{F'}]$ is also connected.

To show the claim, we first note that $C(e) \cap C(e')$ is a strip with width 4. Since $F \cup F'$ is connected, there is a vertex x of $F \cup F'$ in $C(e) \cap C(e')$. x must belong to $F \cap F'$. Let $y \in D_e^F$ dominate x and $y' \in D_{e'}^{F'}$ dominate x. We next consider two cases.

Case 1. $y \in C(e')$ or $y' \in C(e)$. If $y \in C(e')$, then y must belong to F' and hence y is dominated by $D_{e'}^{F'}$. Thus, $G[D_e^F \cup D_{e'}^{F'}]$ is connected.

Case 2. $y \notin C(e')$ and $y' \notin C(e)$. In this case, path (y, x, y') passes through $C(e) \cap C(e')$. However, $d(y, y') \leq 2$. Hence, it is impossible for this case to occur.

The truth of our claim implies that for any connected component F of $G[C(e) \cup C(e')]$, $D_e \cup D_{e'}$ has a connected component dominating F. Putting cells together into a horizontal trip and then putting all horizontal strips together into $P(0)$, we would obtain a property of $D(0)$ that for every connected component of G, $D(0)$ has a connected component to dominate it. Since G is connected, $D(0)$ contains a CDS.

Note that each vertex may appear in $CB(e)$ for at most four cells e. Therefore, by Lemma 3.2.2, $D(0)$ can be computed in time

$$\sum_{e \in P(0)} n_e^{O(a^2)} \leq (4n)^{O(a^2)} = n^{O(a^2)},$$

where $n = |V|$. □

To estimate $|D(0)|$, we consider a minimum solution D^* of MIN-CDS. Let $PB(0) = \cup_{e \in P(0)} B(e)$.

Lemma 3.2.4. *Let $PB(0) = \cup_{e \in P(0)} B(e)$. Then $|D(0)| \leq |D^*| + 24|D^* \cap PB(0)|$.*

Proof. For each cell e, we modify $D^* \cap CB(e)$ into a feasible solution of LOCAL(e) as follows. Consider a connected component F of $G[V \cap C(e)]$. Suppose F is dominated by k connected components $C_1, C_2, \ldots, C_k$ of $G[D^* \cap CB(e)]$ ($k \geq 2$) and they are connected outside of $CB(e)$. Then every C_i has a vertex lying in $CB(e) - C(e)$. Since F is connected, there exist C_i and C_j ($i \neq j$) such that C_i and C_j can be connected together by adding two new vertices. We can charge these two vertices to the one vertex of C_i lying in $CB(e) - C(e)$. Moreover, each vertex x in $D^* \cap (CB(e) - C(e))$ can dominate at most three connected components of $G[V \cap C(e)]$ (this is because each connected component F_i contributes a vertex u_i in a half disk with center at x and radius one, and $d(u_i, u_j) > 1$ for $i \neq j$, which

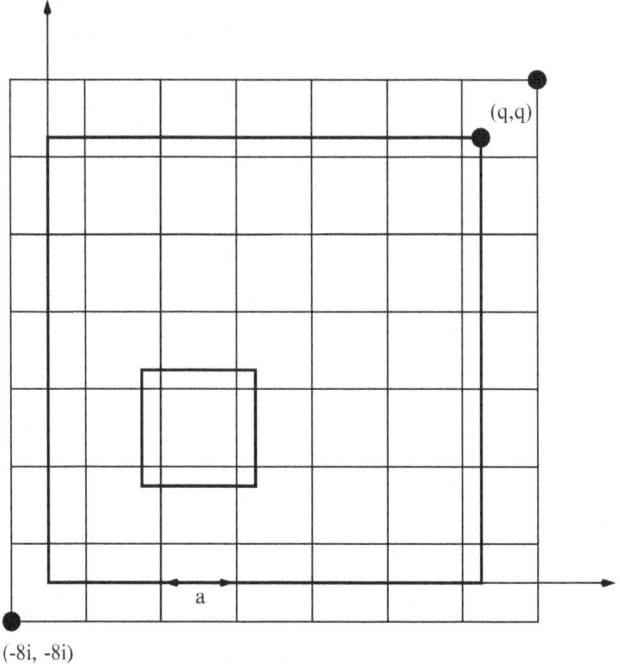

Fig. 3.9 Grid $P(i)$

implies that $\angle u_i x u_j > 60°$). Hence x can be charged at most six times. Let D'_e be obtained from $D^* \cap CB(e)$ by above modification. Then

$$|D'_e| \leq |D^* \cap CB(e)| + 6|D^* \cap (CB(e) - C(e))|$$
$$\leq |D^* \cap e| + 6|D^* \cap B(e)|.$$

Now, note that each vertex can appear in $B(e)$ for at most four cells e and $|D_e| \leq |D'_e|$ where D_e is a minimum solution of LOCAL(e). Thus, we have

$$|D(0)| \leq \sum_{e \in P(0)} |D'_e|$$
$$\leq \sum_{e \in P(0)} (|D^* \cap e| + 6|D^* \cap B(e)|)$$
$$\leq |D^*| + 24|D^* \cap PB(0)|. \qquad \square$$

Now, we shift partition $P(0)$ to $P(i)$ as shown in Fig. 3.9 such that the left and lower corner of the grid is moved to point $(-8i, -8i)$. For each $P(i)$, we can compute a feasible solution $D(i)$ in the same way as computing $D(0)$ for $P(0)$. Then we have

3.2 NP-Hardness and PTAS

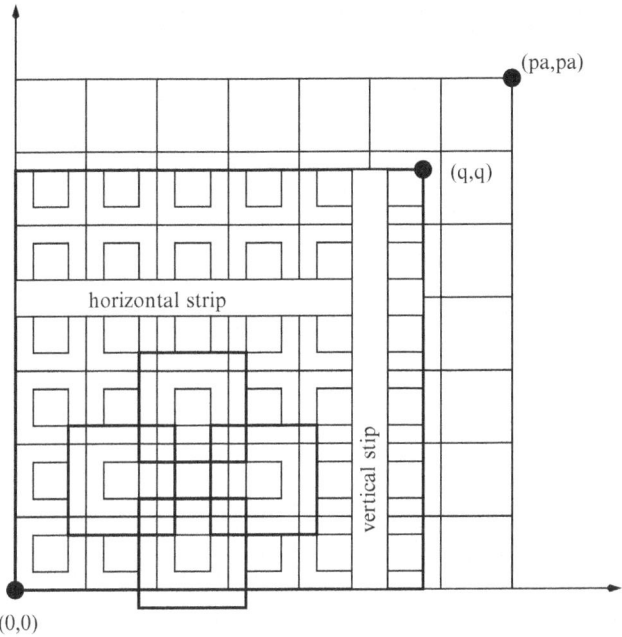

Fig. 3.10 Horizontal and vertical strips

(a) $D(i)$ is a CDS.
(b) $D(i)$ can be computed in time $n^{O(a^2)}$.
(c) $|D(i)| \leq |D^*| + 24|D^* \cap PB(i)|$.

From (c), we can obtain the following

Lemma 3.2.5. *For* $k = a/8$, $|D(0)| + |D(1)| + \cdots + |D(k-1)| \leq (k+48)|D^*|$.

Proof. Note that $PB(i)$ consists of a group of horizontal strips and a group of vertical strips (Fig. 3.10). All horizontal strips in $PB(0) \cup PB(1) \cup \cdots \cup PB(k-1)$ are disjoint and all vertical strips in $PB(0) \cup PB(1) \cup \cdots \cup PB(k-1)$ are also disjoint. Therefore,

$$\sum_{i=0}^{k-1} |D^* \cap PB(i)| \leq 2|D^*|.$$

Hence,

$$\sum_{i=0}^{k-1} |D(i)| \leq (k+48)|D^*|. \qquad \square$$

Set $k = \lceil 1/(8\varepsilon) \rceil$ and run the following algorithm.

Algorithm PTAS
input a unit disk graph G;
  Compute $D(0), D(1), \ldots, D(k-1)$;
  Choose i^*, $0 \le i^* \le k-1$ such that
    $|D(i^*)| = \min(|D(0)|, |D(1)|, \ldots, |D(k-1)|)$;
output $D(i^*)$.

Theorem 3.2.6 (Cheng et al. [22]). *Algorithm PTAS produces an approximation solution for* MIN-CDS *with size*

$$|D(i^*)| \le (1+\varepsilon)|D^*|$$

and runs in time $n^{O(1/\varepsilon^2)}$.

Proof. It follows from Lemmas 3.2.3 and 3.2.5. □

3.3 Two-Stage Algorithm

Although MIN-CDS in unit disk graph has a PTAS, the running time is a polynomial with very high degree and hence not able to be implemented for a real world problem. Therefore, one still wants to find faster approximations with small constant performance ratio. So far, all approximation algorithms of this type are designed in the same manner: First, construct a maximal independent set and then connected it into a CDS. Here, one notes that every maximal independent set is a dominating set.

To save the spending at the second stage, one usually constructs a maximal independent set in the following way:

Algorithm MIS
input a connected graph;
Color a node in black, its neighbors in grey and all other nodes in white;
while a white node exists **do**
    choose a white node x with a grey neighbor and
    color x in black and its white neighbors in grey;
output the set of black nodes.

The maximal independent set constructed as above has the following property.

Lemma 3.3.1 (AoA Property). *Every subset of the maximal independent set constructed as above is within distance two from its complement.*

In the second stage, consider the constructed maximal independent set as a set of terminals and then find the minimum number of Steiner nodes (added nodes) to interconnect all terminal. This means to solve the following problem.

3.3 Two-Stage Algorithm

ST-MSP-IN-UDG: Given a unit disk graph $G = (V,E)$ and a node subset $P \subseteq V$ with AoA Property, find a node subset S with the minimum cardinality, such that $G[P \cup S]$ is connected. (Nodes in P are called *terminals* while nodes in S are called *Steiner nodes*.)

This is an NP-hard problem with many approximation solutions. Any one of them can play the role in the second stage. The following is an example, a greedy approximation.

For any subset C of nodes, let $p(C)$ denote the number of connected components of $G[C]$. Denote $\Delta_x p(C) = p(C \cup \{x\}) - p(C)$. Suppose a maximal independent set D with AoA property is already constructed.

Greedy Connection
input a dominating set D;
$C \leftarrow D$;
while $p(C) \geq 2$ **do**
    choose a node x to maximize $-\Delta_x p(C)$ and
    $C \leftarrow C \cup \{x\}$;
output C.

The following theorem states the performance of this approximation.

Theorem 3.3.2 (Zou et al. [132]). *Suppose G is a graph with $\alpha(G) \leq a \cdot \gamma_c(G) + b$ and D is a maximal independent set with AoA property. Then the CDS produced by* **Greedy Connection** *has size at most*

$$(a + 2 + \ln(a-1))\gamma_c(G) + b + \lfloor b \rfloor - 1.$$

Proof. Suppose $x_1, \ldots, x_g$ are selected in turn by the greedy algorithm. Let $\{y_1, \ldots, y_{\gamma_c(G)}\}$ be a minimum CDS and for any i, $\{y_1, \ldots, y_i\}$ induces a connected subgraph. Denote $C_0 = D$, $C_{i+1} = C_i \cup \{x_{i+1}\}$ and $C_j^* = \{y_1, \ldots, y_j\}$. Then

$$-\Delta_{y_j} p(C_i \cup C_{j-1}^*) + \Delta_{y_j} p(C_i) \leq 1.$$

So, $-\Delta_{x_{i+1}} p(C_i) \geq -\Delta_{y_j} p(C_i)$ for all $1 \leq j \leq \gamma_c(G)$. Thus,

$$-\Delta_{x_{i+1}} p(C_i) \geq \frac{-\sum_{j=1}^{\gamma_c(G)} \Delta_{y_j} p(C_i)}{\gamma_c(G)}$$

$$\geq \frac{-\gamma_c(G) + 1 - \sum_{j=1}^{\gamma_c(G)} \Delta_{y_j} p(C_i \cup C_{j-1}^*)}{\gamma_c(G)}$$

$$= \frac{-\gamma_c(G) + 1 - p(C_i \cup C^*) + p(C_i)}{\gamma_c(G)}$$

$$= \frac{\gamma_c(G) + p(C_i)}{\gamma_c(G)},$$

that is,
$$-p(C_{i+1}) \geq -p(C_i) + \frac{-\gamma_c(G) + p(C_i)}{\gamma_c(G)}.$$

Denote $a_i = -\gamma_c(G) - b + p(C_i)$. Then
$$a_{i+1} \leq a_i \left(1 - \frac{1}{\gamma_c(G)}\right).$$

Thus,
$$a_i \leq a_0 \left(1 - \frac{1}{\gamma_c(G)}\right)^i \leq a_0 e^{-i/\gamma_c(G)}.$$

First, assume the existence of i, $0 \leq i < g$ such that
$$a_{i+1} < \gamma_c(G) \leq a_i.$$

Then $g \leq i + 2\gamma_c(G) - 1 + \lfloor b \rfloor$ and
$$\gamma_c(G) \leq a_0 e^{-i/\gamma_c(G)}.$$

Hence,
$$i \leq \gamma_c(G) \ln(a_0/\gamma_c(G)).$$

Moreover,
$$a_0/\gamma_c(G) = (-\gamma_c(G) - b + |D|)/\gamma_c(G) \leq a - 1.$$

Therefore,
$$|D| + g \leq (a + 2 + \ln(a-1))\gamma_c(G) + b + \lfloor b \rfloor - 1.$$

Now, consider the case that there is no i such that $a_{i+1} < \gamma_c(G) \leq a_i$. Note that $a_g = -\gamma_c(G) - b + 1 < \gamma_c(G)$. Thus, it must have $a_0 < \gamma_c(G)$. This implies that $g \leq 2\gamma_c(G) - 2 + \lfloor b \rfloor$. Thus,
$$|D| + g \leq (a + 2)\gamma_c(G) + b + \lfloor b \rfloor - 2. \qquad \square$$

There is a better analysis found by Wan et al. [106]. They found some geometric properties of this approximation and gave a better performance ratio by taking this advantage.

The best-known approximation for ST-MSP-IN-UDG is given by Min et al. [79] as follows.

Algorithm MHHW:
input a maximal independent set with AoA property.

Color all its nodes in black and others in gray. In the following, we will change some gray nodes to black in certain rules. A *black component* is a connected component of the subgraph induced by black nodes.

Stage 1 **while** there exists a grey node adjacent to at least three
　　　　　　black components **do**
　　　　　　　　change its color from gray to black;
　　end-while;
Stage 2 **while** there exists a grey node adjacent to at least two
　　　　　　black components **do**
　　　　　　　　change its color from gray to black;
　　end-while;
　　output all black nodes.

They showed the following.

Theorem 3.3.3 (Min et al. [79]). *In Algorithm MHHW, the number of gray nodes changed their color to black is at most* $3 \cdot \gamma_c(G)$ *where G is input unit disk graph.*

3.4 Independent Number (I)

Two points u and v are *independent* if $d(u,v) > 1$. To establish the upper bound of independent number $\alpha(G)$ for unit disk graphs G, one way is to study packing independent points in the neighborhood area of the minimum CDS.

The following result is first proved by Wan et al. [104].

Lemma 3.4.1 (Wan et al. [104]). *A disk D with radius one can contain at most five independent points.*

Proof. Let o be the center of D. Suppose $u_1, u_2, \ldots, u_k$ are all independent points in D, in counterclockwise ordering. Then we must have $\angle u_1 x u_2 > 60°$, $\angle u_2 x u_3 > 60°, \ldots, \angle u_k x u_1 > 60°$ since $d(x, u_i) \leq 1$ and $d(u_1, u_2) > 1, d(u_2, u_3) > 1, \ldots, d(u_k, u_1) > 1$. Therefore, $k \cdot 60° < 360°$. Hence, $k \leq 5$. □

With Lemma 3.4.1, Wan et al. [104] proved the following

Theorem 3.4.2 (Wan et al. [104]). *Let $\alpha(G)$ and $\gamma_c(G)$ be the independent number and the connected dominating number of unit disk graph G, respectively. Then*

$$\alpha(G) \leq 4 \cdot \gamma_c(G) + 1.$$

Proof. The proof is by induction on $\gamma_c(G)$. If $\gamma_c(G) = 1$, then the inequality follows immediately from Lemma 3.4.1. In general, suppose $\gamma_c(G) = n > 1$. Choose a node x in the minimum CDS C such that $C - \{x\}$ is still connected. This can be done by choosing x as a leaf of spanning tree of $G[C]$. By induction hypothesis, there are at most $4(n-1) + 1$ independent points lying in $\mathcal{A} = \cup_{y \in C - \{x\}} \text{disk}_1(y)$.

Let $z \in C - \{x\}$ be adjacent to x. Suppose $w_1, \ldots, w_k$ are independent points in $\mathrm{disk}_1(x) \setminus \mathcal{A}$. Note that every point in $\mathrm{disk}_1(x) \setminus \mathcal{A}$ is independent from z. Thus, $z, w_1, \ldots, w_k$ are independent in $\mathrm{disk}_1(x)$. By Lemma 3.4.1, $k \leq 4$. Therefore, there exist at most $4(n-1) + 1 + 4 = 4n + 1$ independent points lying in $\mathcal{A} \cup \mathrm{disk}_1(x)$. □

Wu et al. [123] showed a result on packing independent points in two disks.

Lemma 3.4.3 (Wu et al. [123]). *Let u and v be two points with distance at most one. Then $\mathrm{disk}_1(u) \cup \mathrm{disk}_1(v)$ can contain at most eight independent points.*

Proof. For contradiction, suppose there exists an independent set I of at least nine points lying in $\mathrm{disk}_1(u) \cup \mathrm{disk}_1(v)$. One claims that the intersection $A = \mathrm{disk}_1(u) \cap \mathrm{disk}_1(v)$ contains at most one point in I.

Indeed, suppose A contains k vertices in I. By Lemma 3.4.1, $\mathrm{disk}_1(u) - A$ contains at most $5-k$ points in I and $\mathrm{disk}_1(v) - A$ contains at most $5-k$ points in I. Thus, $\mathrm{disk}_1(u) \cup \mathrm{disk}_1(v)$ contains at most $10-k$ point in I. Hence, $10-k \geq 9$, that is, $k \leq 1$.

In Lemma 3.4.4, one shows that $\mathrm{disk}_1(u) \cup \mathrm{disk}_1(v) - A$ contains at most seven independent points. Therefore, $\mathrm{disk}_1(u) \cup \mathrm{disk}_1(v)$ contains at most eight independent points, a contradiction. □

Lemma 3.4.4. *Let u and v be two points with distance at most one. Then $\mathrm{disk}_1(u) \triangle \mathrm{disk}_1(v)$ can contain at most seven independent points where*

$$\mathrm{disk}_1(u) \triangle \mathrm{disk}_1(v) = (\mathrm{disk}_1(u) \setminus \mathrm{disk}_1(v)) \cup (\mathrm{disk}_1(v) \setminus \mathrm{disk}_1(u)).$$

Proof. By Lemma 3.4.1, $\mathrm{disk}_1(u) \setminus \mathrm{disk}_1(v)$ contains at most four independent points and $\mathrm{disk}_1(v) \setminus \mathrm{disk}_1(u)$ contains at most four independent points. For contraction, suppose $\mathrm{disk}_1(u) \triangle \mathrm{disk}_1(v)$ contains eight independent points. Then $\mathrm{disk}_1(u) \setminus \mathrm{disk}_1(v)$ contains exactly four independent points a_1, a_2, a_3, a_4 and $\mathrm{disk}_1(v) \setminus \mathrm{disk}_1(u)$ contains exactly four independent points a_5, a_6, a_7, a_8. Assume $a_1, \ldots, a_4$ lie counter-clockwisely in $\mathrm{disk}_1(u)$ and $a_5, \ldots, a_8$ lie counter-clockwisely in $\mathrm{disk}_1(v)$. Denote by ub_i the radius passing through a_i for $i = 2, \ldots, 4$ and by vb_i the radius passing through a_i for $i = 5, \ldots, 8$. Using arcs with radius one, draw four arc-triangles ub_2c_2, ub_3c_3, vb_6c_6, and vb_7c_7 as shown in Fig. 3.11. Their boundaries intersect the boundary of $\mathrm{disk}_1(u) \cap \mathrm{disk}_1(v)$ at d_2, d_3, d_6, d_7, respectively. Note that none of a_1, a_4, a_5, a_8 can lie in the four arc-triangles ub_2c_2, ub_3c_3, vb_6c_6, and vb_7c_7. Therefore, a_1, a_4, a_5, a_8 must lie in the four small dark areas xc_2d_2, yc_3d_3, yc_6d_6 and xc_7d_7, respectively, as shown in Fig. 3.11.

Next, one find a contradiction by proving the fact that there exist two small dark areas too close to contain two independent vertices.

To show this fact, note that $\angle b_2ub_3 > 60°$ and $\angle c_2ub_2 = \angle b_3uc_3 = 60°$. Hence, $\angle c_2uc_3 > 180°$ and $\angle c_3uc_2 < 180°$ (note that $\angle c_3uc_2$ is the one obtained by moving c_3u counterclockwise to c_2u). Similarly, $\angle c_7vc_6 < 180°$. Therefore $\angle uc_2c_7 + \angle c_2c_7v + \angle vc_6c_3 + \angle c_6c_3u > 360°$. This means that either $\angle uc_2c_7 + \angle c_2c_7v > 180°$

3.4 Independent Number (I)

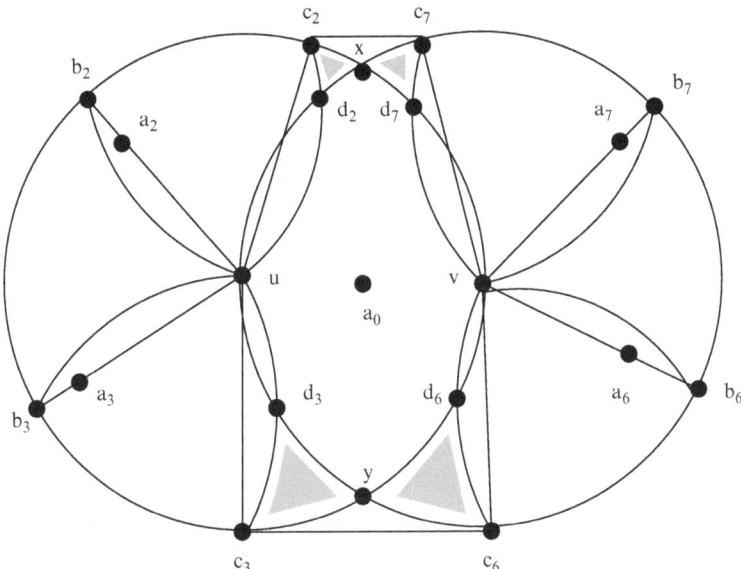

Fig. 3.11 Four small *dark* areas

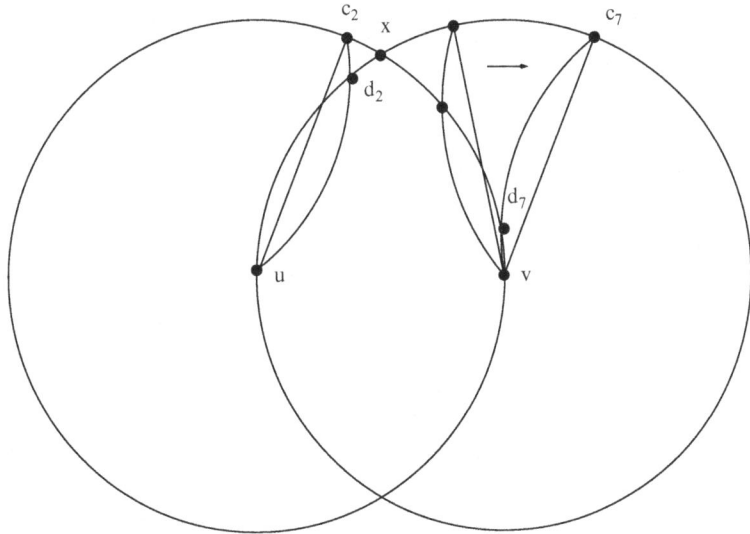

Fig. 3.12 Turn unit arc-triangle vb_7c_7 until $vc_7 \parallel uc_2$

or $\angle vc_6c_3 + \angle c_6c_3u > 180°$. Without loss of generality, assume the former occurs (Fig. 3.12). Next, it will be showed that dark areas xc_2d_2 and xc_7d_7 cannot contain two independent points.

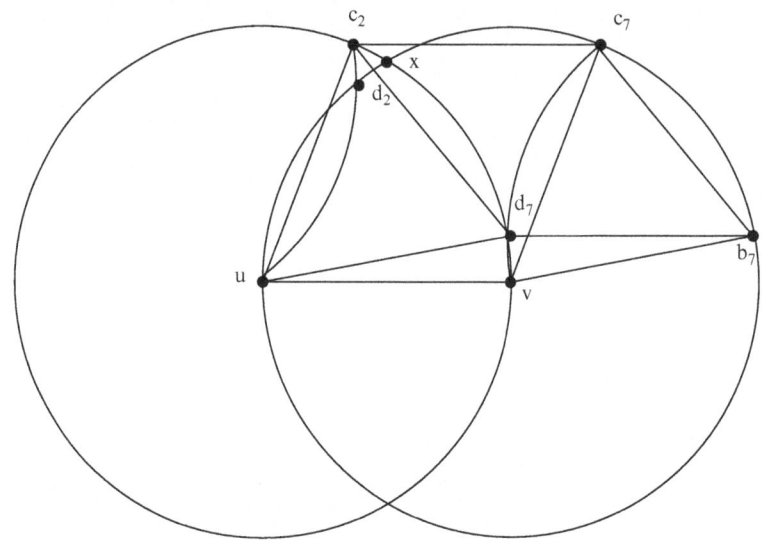

Fig. 3.13 Move u until $|uv| = 1$

To do so, at the first, the area xc_7d_7 is enlarged by turning the arc-triangle vb_7c_7 around v until vc_7 is parallel to uc_2. At this limit position, quadrilateral c_2uvc_7 becomes a parallelogram so that $|c_2c_7| = |uv| \leq 1$. It follows that the distance between two points in areas xc_2d_2 and xc_7d_7 cannot exceed $\max(|c_2c_7|, |c_2d_7|, |d_2c_7|, |d_2d_7|)$. Moreover, it can be proved that $|d_2d_7| \leq \max(|c_2d_7|, |d_2c_7|)$. In fact, note that $\angle c_7d_7d_2 + \angle d_7d_2c_2 > 360°$. Thus, either $\angle c_7d_7d_2 > 90°$ or $\angle d_7d_2c_2 > 90°$. Therefore, either $|d_2c_7 > |d_2d_7|$ or $c_2d_7 > |d_2d_7|$.

Now, to complete the proof of claimed fact, it suffices to prove that $|c_2d_7| \leq 1$ and $|d_2c_7| \leq 1$. To see $|c_2d_7| \leq 1$, make $|c_2d_7|$ longer by moving v away from u until $|uv| = 1$ (Fig. 3.13). At this limit position, $|uv| = |vb_7| = |b_7d_7| = |d_7u| = 1$. Therefore, uvb_7d_7 is a parallelogram. It follows that $|d_7b_7| = |c_2c_7| = 1$ and d_7b_7 is parallel to uv and hence parallel to c_2c_7. Thus, $c_2d_7b_7c_7$ is a parallelogram. Therefore, $|c_2d_7| = |c_7b_7| = 1$. Similarly, one can show $|d_2c_7| \leq 1$. □

With Lemma 3.4.3, Wu et al. [123] established the following

Theorem 3.4.5 (Wu et al. [123]). *Let $\alpha(G)$ and $\gamma_c(G)$ be the independent number and the connected dominating number of unit disk graph G, respectively. Then*

$$\alpha(G) \leq 3.8 \cdot \gamma_c(G) + 1.2.$$

Proof. First, one proves the following two lemmas about unit disk graph and general graphs.

Lemma 3.4.6. *In any unit disk graph, there exists a minimum spanning tree such that every vertex has degree at most five.*

3.4 Independent Number (I)

Proof. First, note that in any minimum spanning tree, each vertex u has degree at most six. In fact, for contradiction, suppose u has degree more than six. Then there are two edges uv and uv' such that $\angle vuv' < 60°$. It follows that $|vv'| < |uv|$ or $|vv'| < |uv'|$. Replacing uv (in the former case) or uv' (in the latter case) by vv' would result in a shorter spanning tree, a contradiction. A similar argument can also proved that if a vertex u has degree six, then all edges at u have the same length and all angles at u equal $60°$.

Suppose T is a minimum spanning tree with the minimum number of vertices with degree six. For contradiction, suppose T has a vertex u with degree six. Then, every angle at u equals $60°$ and all edges incident to u have the same length. Consider a vertex v adjacent to u. Then u has two edges uw and ux such that $\angle wuv = \angle vux = 60°$ and $|uv| = |uw| = |ux|$. Thus, $|vw| = |uw|$ and $|vx| = |ux|$. Replacing uw and ux by vw and vx results in still a minimum spanning tree. But, v gets two more edges. This means that v has degree at most four in T. Thus, replacing uv by vw in T would result in a minimum spanning tree in which u has degree five and v has degree at most five, so that the number of vertices with degree six is reduced by one, a contradiction. □

Lemma 3.4.7. *Every tree T with at least three vertices has a non-leaf vertex adjacent to at most one non-leaf vertex.*

Proof. Let T' be the subtree obtained from T by removal of all leaves. Since T has at least three vertices, T' contains at least one vertex. If T' contains only one vertex, then it meets our requirement. If T' contains more than one vertex, then every leaf of T' is a non-leaf vertex of T satisfying the condition stated in the lemma. □

Now, it is ready to prove Theorem 3.4.5. Let H be a subgraph induced by a minimum CDS in the given unit disk graph G. Then H is also a unit disk subgraph. By Lemma 3.4.6, H has a minimum spanning tree T such that every vertex has degree at most five. Let $|T|$ denote the number of vertices in T. It will be proved by induction on $|T|$ that there exist at most $3.8|T| + 1.2$ independent vertices in the neighbor area of T. For $|T| = 1$ or 2, this is true by Lemmas 3.4.1 and 3.4.3. Next, assume $|T| \geq 3$. By Lemma 3.4.7, T contains a non-leaf vertex v adjacent to at most one non-leaf vertex. Let u be the non-leaf neighbor of v if it exists, or a leaf neighbor of v, otherwise. Let $x_1, \ldots, x_k$ ($k \leq 4$) be other neighbors of v. Note that by Lemma 3.4.1, each $disk_1(x_i)$ for $1 \leq i \leq k-1$ contains at most four independent points which are also independent from v, and by Lemma 3.4.3, $disk_1(v) \cup disk_1(x_k)$ contains at most seven independent points which are also independent from u. Moreover, by the induction hypothesis, the neighbor area of $T - \{v, x_1, \ldots, x_k\}$ contains at most $3.8(|T| - k - 1) + 1.2$ independent vertices. Therefore, the neighbor area of T contains at most

$$3.8(|T| - k - 1) + 1.2 + 7 + 4(k-1) = 3.8|T| + 1.2 + 0.2(k-4) \leq 3.8|T| + 1.2$$

independent vertices. Note that $|T| = mcds$. This completes the proof of Theorem 3.4.5. □

Wan, Wang and Yao [106] found an idea to prove a better bound based on the study on packing independent points in the neighborhood area of a star.

First, they note that every tree can be partitioned into nontrivial stars. (A star is trivial if it contains only one node.)

Lemma 3.4.8. *For any tree T, its node set has a partition $V(T) = (V_1,\ldots,V_k)$ such that for every part V_i, $T[V_i]$ is a star with at least two nodes.*

Proof. Choose any node r of T as a root and consider T as a rooted tree. Then one can compute such a partition as follows.

$V \leftarrow V(T)$;
$i \leftarrow 0$;
while $V \neq \emptyset$ **do begin**
    $i \leftarrow i+1$;
    choose a leaf u at lowest level and find its parent node v;
    let V' be the set of v and its all children;
    if $|V - V'| > 1$
      then $V_i \leftarrow V'$
      else $V_i \leftarrow V$
    $V \leftarrow V - V_i$;
end-while
output $(V_1,\ldots,V_i)$.

□

Wan et al. [106] then found tight upper bound for the number of independent points lying in the neighborhood area of a star.

Lemma 3.4.9. *The neighborhood area of a star with n nodes can contain at most ϕ_n independent points where*

$$\phi_n = \begin{cases} 3n+2 & \text{if } n \leq 2, \\ 3n+3 & \text{if } n \leq 5, \\ 21 & \text{if } 6 \leq n. \end{cases}$$

Consider a star S. Let o be the center of S. Then the neighborhood area of S is contained in $\text{disk}_2(o)$. By Zassenhaus–Groemer–Oler inequality (in Section 3.6), $\text{disk}_2(o)$ can contain at most 21 independent points. This means that Lemma 3.4.9 holds for $n \geq 6$. By Lemmas 3.4.1 and 3.4.3, Lemma 3.4.9 holds for $n \leq 2$. For $n = 3$, suppose $V(S) = \{v_1, v_2, v_3\}$. By Lemma 3.4.3, $\text{disk}_1(v_1) \cup \text{disk}_1(v_2)$ can contain at most eight independent points and by Lemma 3.4.1, $\text{disk}_1(v_3) \setminus (\text{disk}_1(v_1) \cup \text{disk}_1(v_2))$ can contain at most four independent points. Thus, $\text{disk}_1(v_1) \cup \text{disk}_1(v_2) \cup \text{disk}_1(v_3)$ can contain at most twelve independent points. For $n = 4, 5$, the proof is given by a tedious geometric argument and the interested reader may see the original paper [106] for detail.

3.4 Independent Number (I)

Theorem 3.4.6 (Wan et al. [106]). *For any connected unit disk graph G with at least two nodes,*

$$\alpha(G) \leq \frac{11}{3} \cdot \gamma_c(G) + 1.$$

Proof. Let $S = \{S_1, \ldots, S_k\}$ be a nontrivial star partition of a spanning tree of the minimum CDS of G such that for any $1 \leq i \leq k$, $S_1 \cup \cdots \cup S_i$ is connected. Let I be the maximum independent set of G. For any subgraph H, denote by $I(H)$ the intersection of I and the neighborhood area of H. The proof is by a mathematical induction on k. By Lemma 3.4.9,

$$|I(S_k)| \leq \frac{11}{3} \cdot |S_k| + 1.$$

Since $\cup_{i=1}^{k} S_i$ is connected, S_k must have a node v lying in the neighborhood area of $\cup_{i=1}^{k-1} S_i$ and v is independent to any point in $I(\cup_{i=1}^{k-1} S_i) \setminus I(S_k)$. By the induction hypothesis,

$$|I(\cup_{i=1}^{k-1} S_i) \setminus I(S_k)| + 1 \leq \frac{11}{3} \cdot |\cup_{i=1}^{k-1} S_i| + 1,$$

that is,

$$|I(\cup_{i=1}^{k-1} S_i) \setminus I(S_k)| \leq \frac{11}{3} \cdot |\cup_{i=1}^{k-1} S_i|.$$

Therefore,

$$|I(\cup_{i=1}^{k} S_i)| = |I(\cup_{i=1}^{k-1} S_i) \setminus I(S_k)| + |I(S_k)|$$

$$\leq \frac{11}{3} \cdot |\cup_{i=1}^{k} S_i| + 1. \qquad \square$$

Vahdatpour et al. [103] claimed that they proved that for any connected unit disk graph G,

$$\alpha(G) \leq 3\gamma_c(G) + 3.$$

If their proof is correct, then this is the best possible result. Wan et al. [106] have showed that for some unit disk graph G

$$\alpha(G) = \gamma_c(G) + 3.$$

Unfortunately, the proof of Vahdatpour et al. is far from a complete one. In the following, we give an analysis on their proof and indicate what important parts their proof miss. First, note that their proof use a mathematical induction on the number of vertices in the minimum CDS based on two important lemmas.

Let T be a spanning tree of the minimum CDS. For any node v, denote $N(v) = disk_1(v)$. Let U be any set of independent points lying in the neighborhood area $Q(T)$ of T. Assume $v_1, v_2, \ldots, v_T$ is an arbitrary traversal of T. For any i, $2 \leq i \leq |T|$, consider $U_i = N(v_i) \cap U - \cup_{j=1}^{i-1} N(v_j)$ be the subset of nodes in U that are adjacent to v_i but not to any of $v_1, v_2, \ldots, v_{i-1}$. We will call U_i the semi-exclusive neighboring set of node v_i." The first lemma is as follows:

Lemma 3.4.10. *For two distinct vertices v_i and v_j of T with $|U_i| = |U_j| = 4$, there exists a node v_k on the path between v_i and v_j such that $|U_k| \leq 2$.*

Now, consider a leaf v_j. There are several cases.

Case 1. $|U_j| \leq 3$. Then we apply the induction hypothesis on $T \setminus v_j$ and finish the induction proof.

Case 2. $|U_j| = 4$ and $|U_i| \leq 3$ for every $i \neq j$. In this case, we immediately have $|U| \leq 3|T| + 3$.

Case 3. $|U_j| = 4$ and there is $i \neq j$ such that $|U_i| = 4$. By Lemma 3.4.10, there exists a vertex v_k on the path between v_i and v_j such that $|U_k| \leq 2$.

Subcase 3.1. Path $P = (v_j, \ldots, v_k)$ does not contain a fork vertex (i.e., a vertex with degree at least three). In this subcase, we can apply the induction hypothesis to $T \setminus P$ and finish the induction proof.

Subcase 3.2. Path $P = (v_j, \ldots, v_k)$ contains some fork vertices. In this subcase, $T \setminus P$ is not connected and hence not a tree so that we cannot apply induction hypothesis to $T \setminus P$. This is a complicated subcase. However, Vahdatpour et al. [103] did not give sufficient argument to deal with it. Indeed, they provided the second lemma to handle this subcase. However, (1) the second lemma is not sufficient to handle this subcase and (2) the proof of the second lemma is far from a complete one.

Therefore, it is still an open problem whether the inequality (3.1) holds or not.

3.5 Independent Number (II)

Funke et al. [51] initiated another idea to establish the upper bound of the independent number $\alpha(G)$ for unit disk graph G. The idea is based on the fact that all $disk_{0.5}(v)$ for v over nodes in the maximum independent set are disjoint and they all lie in the union Ω of $disk_{1.5}(x)$ for x over all nodes in the minimum CDS. The following result follows immediately from this fact.

Theorem 3.5.1 (Funke et al. [51]). *Let $\alpha(G)$ and $\gamma_c(G)$ be the independent number and the connected dominating number of unit disk graph G, respectively. Then*

$$\alpha(G) \leq 3.748 \gamma_c(G) + 5.252.$$

Proof. Define the dominating area of a vertex x to be the disk $disk_{1.5}(x)$. Then two adjacent nodes have at least $\frac{9}{2} \arccos \frac{1}{3} - \sqrt{2}$ area in common. Thus, the union Ω of dominating areas of a minimum CDS can have at most

3.5 Independent Number (II)

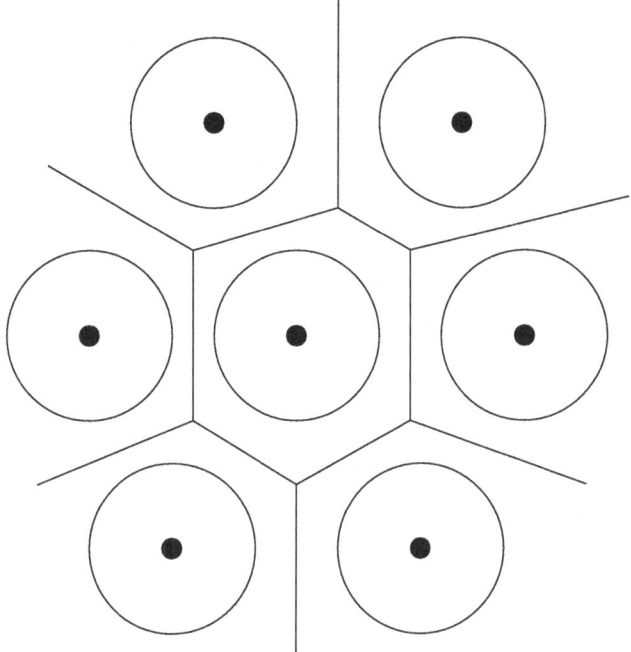

Fig. 3.14 Voronoi division

$$(\gamma_c(G) - 1)\left(\frac{9}{2}\arccos\frac{1}{3} - \sqrt{2}\right) + \pi 1.5^2$$

area. For every node v in a maximal independent set, draw a disk $\text{disk}_{0.5}(v)$. All such disks are disjoint and lie in the adjacent area of the maximum independent set. Therefore, the size of a maximal independent set $\alpha(G)$ is at most

$$\frac{(\gamma_c(G) - 1)(\frac{9}{2}\arccos\frac{1}{3} - \sqrt{2}) + \pi 1.5^2}{0.25\pi} \leq 3.748\gamma_c(G) + 5.252. \qquad \square$$

To improve this approach, Funke et al. further introduced Voronoi division (Fig. 3.14) of the maximum independent set. Denote by $\text{voro}(v)$ the Voronoi cell of node v. Since $\text{disk}_{0.5}(v) \subset \text{voro}(v)$, the area of $\text{voro}(v)$ would be bigger that the area of $\text{disk}_{0.5}(v)$. In fact, they claimed that the area of $\text{voro}(v)$ is at least $\sqrt{3}/2$ and $\text{voro}(v) \cap \Omega$ has area at least 0.8525. This fact has been verified by Gao et al. [54]. Therefore, Funde et al. [51] established the following.

Theorem 3.5.2 (Funke et al. [51]). *Let $\alpha(G)$ and $\gamma_c(G)$ be the independent number and the connected dominating number of unit disk graph G, respectively. Then*

$$\alpha(G) \leq 3.453\gamma_c(G) + 4.839.$$

Proof. Similar to the proof of Theorem 3.5.1.

Fig. 3.15 The proof of Lemma 3.5.3

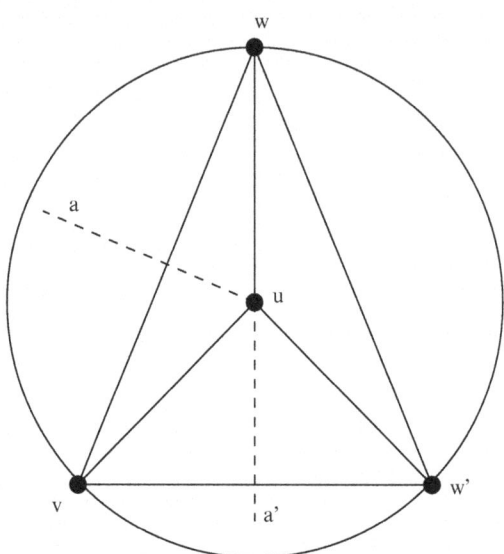

$$\alpha(G) \leq \frac{(\gamma_c(G)-1)(\frac{9}{2}\arccos\frac{1}{3}-\sqrt{2})+\pi 1.5^2}{0.8525} \leq 3.453\gamma_c(G)+4.839. \qquad \square$$

Li et al. [72] found two ideas to make an improvement. The first idea is based on the following facts.

Lemma 3.5.3. *Every vertex u of Voronoi cell* voro(v) *lies outside the disk* disk$_{1/\sqrt{3}}(v)$.

Proof. Suppose ua and ua' are two edges of voro(v) at vertex u. Let w be the symmetric point of v with respect to line ua and w' the symmetric point of v with respect to line ua' (Fig. 3.15). Then by the construction of Voronoi division, it can be seen that v, w and w' are independent and they are on circle circle$_{d(v,u)}(u)$. Note that one of angles $\angle vuw, \angle wuw', \angle w'uv$ is at most 120°. This means one of $d(v,w), d(w,w'), d(w',v)$ is at most $\sqrt{3} \cdot d(v,u)$, that is,

$$1 < \sqrt{3}d(v,u) \text{ or } d(v,u) > 1/\sqrt{3}. \qquad \square$$

Lemma 3.5.4. *Let P be a polygon inscribed in the circle* circle$_{1/\sqrt{3}}(v)$ *such that* disk$_{0.5}(v) \subset P$. *Then*

$$\text{area}(P) \geq \sqrt{3}/2,$$
$$\text{area}(P \cap \text{disk}_{1.5}(s)) \geq \sigma = 0.85505328..,$$

for disk$_{0.5}(v) \subset$ disk$_{1.5}(s)$.

3.6 Zassenhaus–Groemer–Oler Inequality

By Lemma 3.5.3, for each v in the maximum independent set, one can construct a polygon P_v which is inscribed in circle $\text{circle}_{1/\sqrt{3}}(v)$ and $\text{disk}_{0.5}(v) \subset P_v \subset \text{voro}(v)$. By Lemma 3.5.4, the area of $P_v \cap \Omega \geq \sigma$ and hence $\alpha(G) \leq \text{area}(\Omega)/\sigma$.

The second idea is motivated from an observation on Lemma 3.5.4. Lemma 3.5.4 indicates that the maximum independent set I can be partitioned into two parts

$$I_1 = \{v \mid \text{voro}(v) \cap \text{disk}_{1/\sqrt{3}}(v) \subseteq \Omega\},$$
$$I_2 = \{v \mid \text{voro}(v) \cap \text{disk}_{1/\sqrt{3}}(v) \not\subseteq \Omega\}.$$

For $v \in I_1$, $\text{area}(\text{voro}(v) \cap \Omega) \geq \sqrt{3}/2$ and for $v \in I_2$, $\text{area}(\text{voro}(v) \cap \Omega) \geq \sigma$. If $|I_2|$ can be upper-bounded in some way, the upper bound for $\alpha(G)$ could be improved. In fact, since

$$\text{area}(\Omega) \geq \frac{\sqrt{3}}{2} \cdot |I_1| + \sigma \cdot |I_2|,$$

one has

$$|I| \leq \frac{\text{area}(\Omega)}{\frac{\sqrt{3}}{2}} + \left(1 - \frac{\sigma}{\frac{\sqrt{3}}{2}}\right) \cdot |I_2|.$$

Li et al. successfully established an upper bound for $|I_2|$ as follows.

Lemma 3.5.5. *Let C be a minimum CDS of unit disk graph $G = V, E)$. Define $\Omega' = \cup_{x \in C} \text{disk}_{1.5 - 1/\sqrt{3}}(x)$. Then the boundary length of Ω is at most*

$$2\left(3 - \frac{2}{\sqrt{3}}\right)\left((\gamma_c(G) - 1)\arcsin\frac{1}{3 - \frac{2}{\sqrt{3}}} + \frac{\pi}{2}\right)$$

and at least $2(1 - 1/\sqrt{3})|I_2|$.

With this lemma, they showed the following best-known upper bound for $\alpha(G)$.

Theorem 3.5.6 (Li et al. [72]). *Let $\alpha(G)$ and $\gamma_c(G)$ be the independent number and the connected dominating number of unit disk graph G, respectively. Then*

$$\alpha(G) \leq 3.4305176\gamma_c(G) + 4.8184688.$$

3.6 Zassenhaus–Groemer–Oler Inequality

Suppose a compact convex region C contains centers of n non-overlapping unit disks. Then

$$n \leq \frac{2}{\sqrt{3}}A(C) + \frac{1}{2}P(C) + 1,$$

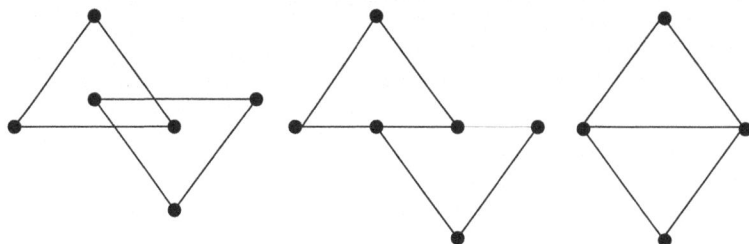

Fig. 3.16 Three families of simplexes

where $A(C)$ is the area of C and $P(C)$ is its perimeter. This inequality is conjectured by Zassenhaus in 1947 (see [125]) and proved independently by Groemer [61] and Oler [85].

This inequality has been used in the proof of Lemma 3.4.9. Indeed, the application of this inequality has been found in many places for analysis of approximation algorithms for optimization problems in unit disk graphs; especially it will be used in later chapters. Therefore, we introduce it here.

There are several proofs of this inequality [50, 61, 78, 85]. The following was given by Folkman and Graham [50] with an extension to two-dimensional simplicial complex.

A zero-dimensional simplex is a point. A one-dimensional simplex is a straight line segment. A two-dimensional simplex is a triangle. In general, a *simplex* is a polytope with minimum number of vertices among all polytopes with certain dimension. For example, a tetrahedron is a three-dimensional polytope with minimum number of vertices and hence a three-dimensional simplex.

Any simplex is the convex hull of its vertices. The convex hull of any subset of vertices in a simplex S is also a simplex, which is called a *face* of simplex S. A family Δ of simplexes is called a *simplicial complex* if it satisfies the following two conditions:

(a) For $S \in \Delta$, every face of S is in Δ.
(b) For $S, S' \in \Delta$, $S \cap S'$ is a face for both S and S'.

From (a) and (b), it is easy to see the following holds:

(c) For $S, S' \in \Delta$, $S \cap S'$ is also a simplex in Δ.

In Fig. 3.16, there are three families of simplexes. While the first two are not simplicial complexes, the last one is.

For any simplex A, $|A|$ denotes the number of vertices in A. Thus, $|A| - 1$ is the dimension of A. The *Euler characteristic* of a simplicial complex Δ is defined by

$$\chi(\Delta) = \sum_{A \in \Delta, A \neq \emptyset} (-1)^{|A|-1} = \sum_{A \in \Delta} (-1)^{|A|-1} + 1.$$

3.6 Zassenhaus–Groemer–Oler Inequality

Let $m(A)$ denote the area of A for two-dimensional simplex A and the length of A for one-dimensional simplex A. For one-dimensional simplex A, let $\varepsilon(A,\Delta)$ denote the number of two-dimensional simplex in Δ having A as a face. For simplicial complex Δ in the Euclidean plane, it is easy to see from (b) that $\varepsilon(A,\Delta) \leq 2$. When $\varepsilon(A,\Delta) = 1$, A is on the boundary of the union of simplexes in Δ. When $\varepsilon(A,\Delta) = 2$, A is in the interior of the union of simplexes in Δ. Now, a proper definition is given for the area $A(\Delta)$ and the perimeter $P(\Delta)$ of a simplicial complex Δ in the Euclidean plane.

$$A(\Delta) = \sum_{A \in \Delta, |A|=3} m(A)$$

and

$$P(\Delta) = \sum_{A \in \Delta, |A|=2} (2 - \varepsilon(A,\Delta)) m(A).$$

The inequality of Folkman and Graham [50] is as follows:

Theorem 3.6.1 (Folkman–Graham [50]). *Let Δ be a simplicial complex in the Euclidean plane. Suppose for any two distinct vertices x and y in Δ, $d(x,y) \geq 1$. Then*

$$|\Delta| \leq \frac{2}{\sqrt{3}} A(\Delta) + \frac{1}{2} P(\Delta) + \chi(\Delta),$$

where $|\Delta|$ is the number of vertices in Δ.

To prove this inequality, the following two lemmas is proved at the first.

Lemma 3.6.2. *Let a, b, c be lengths of three edges of a triangle Δ. Suppose $a \geq b \geq c \geq 1$. Then*

$$\frac{4}{\sqrt{3}} A(\Delta) + a \geq b + c.$$

Proof. By Hero's formula,

$$\begin{aligned}
A(\Delta) &= \frac{1}{4}\sqrt{(a+b+c)(a+b-c)(a-b+c)(-a+b+c)} \\
&\geq \frac{1}{4}\sqrt{(a+b+c)(-a+b+c)} \\
&\geq \frac{1}{4}\sqrt{3c(-a+b+c)} \\
&\geq \frac{1}{4}\sqrt{3(-a+b+c)^2} \\
&= \frac{\sqrt{3}}{4}(-a+b+c).
\end{aligned}$$

Hence,
$$\frac{4}{\sqrt{3}}A(\Delta) + a \geq b + c.$$
□

Lemma 3.6.3. *Let BCDE be a quadrilateral in the Euclidean plane with area A and perimeter P. Suppose the length of every diagonal of BCDE is not less than the length of every edge of BCDE and the length of every edge is at least one. Then*
$$\frac{4}{\sqrt{3}}A - P + 2 \geq 0.$$

Proof. Without loss of generality, assume $\angle B + \angle D \leq \pi$. Note that in this case, one must have $A = A(\triangle BCE) + A(\triangle CDE)$. Since diagonal CE is the longest edge in $\triangle BCE$ and in $\triangle DEC$, one has $\angle B \geq \pi/3$ and $\angle D \geq \pi/3$. Hence, $\pi/3 \leq \angle B \leq 2\pi/3$ and $\pi/3 \leq \angle D \leq 2\pi/3$. Therefore,

$$A = \frac{1}{2}(|BC| \cdot |BE| \cdot \sin\angle B + |DC| \cdot |DE| \cdot \sin\angle D)$$
$$\geq \frac{\sqrt{3}}{4}(|BC| \cdot |BE| + |DC| \cdot |DE|).$$

Thus,
$$\frac{4}{\sqrt{3}}A - P + 2$$
$$\geq |BC| \cdot |BE| + |DC| \cdot |DE| - (|BC| + |BE| + |DC| + |DE|) + 2$$
$$= (|BC| - 1)(|BE| - 1) + (|DC| - 1)(|DE| - 1)$$
$$\geq 0.$$
□

Now, it is ready to prove Theorem 3.6.1.

Proof of Theorem 3.6.1. The proof is an induction on the number of one-dimensional simplexes contained in Δ. First, suppose Δ contains no one-dimensional simplex. Then $A(\Delta) = P(\Delta) = 0$ and $\chi(\Delta)$ is equal to the number of vertices. Therefore, the Folkman–Graham Inequality is true.

Next, suppose Δ contains k one-dimensional simplexes and $k \geq 1$. Assume that for every simplicial complex with less than k one-dimensional complexes, the Folkman–Graham inequality holds. Let $union(\Delta)$ denote the union of simplexes in Δ. Then it is easy to see that for two simplicial complexes Δ and Γ, if $union(\Delta) = union(\Gamma)$, then $A(\Delta) = A(\Gamma)$, $P(\Delta) = P(\Gamma)$ and $\chi(\Delta) = \chi(\Gamma)$. Therefore, it suffices to show the Folkman–Graham Inequality holds for one of simplicial complexes with the same union. Without loss of generality, suppose that Δ has the minimum total length of one-dimensional complexes among those simplicial simplexes with the same union and the same number of one-dimensional simplexes as Δ has.

3.6 Zassenhaus–Groemer–Oler Inequality

Consider a one-dimensional complex σ in Δ with the longest length. There are three cases in the following.

Case 1. $\varepsilon(\sigma,\Delta) = 0$. In this case, $\Delta - \{\sigma\}$ is a simplicial complex and $P(\Delta - \{\sigma\}) = P(\Delta) - 2m(\sigma)$. Therefore, by induction hypothesis,

$$\begin{aligned}
|\Delta| &= |\Delta - \{\sigma\}| \\
&\leq \frac{2}{\sqrt{3}} A(\Delta - \{\sigma\}) + \frac{1}{2} P(\Delta - \{\sigma\}) + \chi(\Delta - \{\sigma\}) \\
&= \frac{2}{\sqrt{3}} A(\Delta) + \frac{1}{2} P(\Delta) - m(\sigma) + \chi(\Delta) + 1 \\
&\leq \frac{2}{\sqrt{3}} A(\Delta) + \frac{1}{2} P(\Delta) + \chi(\Delta).
\end{aligned}$$

Case 2. $\varepsilon(\sigma,\Delta) = 1$. Let τ be the two-dimensional simplex in Δ having σ as a face. Let σ' and σ'' be other two one-dimensional faces of τ. Without loss of generality, assume $m(\sigma') \geq m(\sigma'')$. From the choice of σ, it can be seen that $m(\sigma) \geq m(\sigma') \geq m(\sigma'') \geq 1$. By Lemma 3.6.2,

$$\frac{4}{\sqrt{3}} m(\tau) + m(\sigma) \geq m(\sigma') + \sigma(\sigma'').$$

Note that $\Gamma = \Delta - \{\tau, \sigma\}$ is a simplicial complex. By induction hypothesis,

$$\begin{aligned}
&\frac{2}{\sqrt{3}} A(\Delta) + \frac{1}{2} P(\Delta) + \chi(\Delta) \\
&= \frac{2}{\sqrt{3}} A(\Gamma) + \frac{1}{2} P(\Gamma) + \chi(\Gamma) + \frac{2}{\sqrt{3}} A(\tau) + \frac{1}{2}(m(\sigma) - m(\sigma') - m(\sigma'')) \\
&\geq \frac{2}{\sqrt{3}} A(\Gamma) + \frac{1}{2} P(\Gamma) + \chi(\Gamma) \\
&\geq |\Gamma| = |\Delta|.
\end{aligned}$$

Case 3. $\varepsilon(\sigma,\Delta) = 2$. Let τ and τ' be the two-dimensional simplexes in Δ having σ as a face. Let B and D be two vertices of σ. Suppose C is the third vertex of τ and E is the third vertex of τ'. Because σ is the longest edge in τ and in τ', it can be seen that $\angle BCE \leq \pi/2$, $\angle CEB \leq \pi/2$, $\angle ECD \leq \pi/2$ and $\angle DEC \leq \pi/2$. Thus, $\angle BCD \leq \pi$ and $\angle DEB \leq \pi$. This implies that $BCDE$ is a convex quadrilateral. It follows that a simplicial complex can be obtained from Δ by replacing τ and τ' by $\triangle BCD$ and $\triangle DEB$ with the same union and the same number of one-dimensional simplexes as Δ has. By the choice of Δ, one has $|BD| \geq |CE| = m(\sigma)$. By Lemma 3.6.3,

$$\frac{4}{\sqrt{3}}(A(\tau) + A(\tau')) - P(\tau \cup \tau') + 2 \geq 0.$$

Note that $\Gamma = \Delta - \{\sigma, \tau, \tau'\}$ is a simplicial complex. By induction hypothesis,

$$\frac{2}{\sqrt{3}}A(\Delta) + \frac{1}{2}P(\Delta) + \chi(\Delta)$$

$$= \frac{2}{\sqrt{3}}A(\Gamma) + \frac{1}{2}P(\Gamma) + \chi(\Gamma) + \frac{2}{\sqrt{3}}(A(\tau) + A(\tau')) - P(\tau \cup \tau') + 1$$

$$\geq \frac{2}{\sqrt{3}}A(\Gamma) + \frac{1}{2}P(\Gamma) + \chi(\Gamma)$$

$$\geq |\Gamma| = |\Delta|. \qquad \square$$

Corollary 3.6.4 (Zassenhaus–Groemer–Oler). *Let X be a compact convex region in the Euclidean plane. Suppose X contains a set V of centers of n non-overlapping unit circles. Then*

$$n \leq \frac{2}{\sqrt{3}}A(X) + \frac{1}{2}P(X) + 1,$$

where $A(X)$ and $P(X)$ denote the area and perimeter of X, respectively.

Proof. Let H be the convex hull of V. Then $A(X) \geq A(H)$ and $P(X) \geq P(H)$. Let Δ be a simplicial complex with vertex set V, whose two-dimensional faces form a triangulation of H. Then the union of Δ equals H and $\chi(\Delta) = 1$. By Folkman–Graham Inequality,

$$\frac{2}{\sqrt{3}}A(X) + \frac{1}{2}P(X) + 1 \geq \frac{2}{\sqrt{3}}A(\Delta) + \frac{1}{2}P(\Delta) + 1 \geq |\Delta| = |V|. \qquad \square$$

Corollary 3.6.5. *A disk $\mathrm{disk}_r(o)$ can contain at most*

$$\frac{2}{\sqrt{3}}\pi r^2 + \pi r + 1$$

independent points.

Proof. It follows immediately from Zassenhaus–Graemer–Oler inequality. $\qquad \square$

Chapter 4
CDS in Unit Ball Graphs and Growth Bounded Graphs

> *Time is cubic, not linear as stupid educators teach.*
> GENE RAY

4.1 Motivation and Overview

In a mountain area or underwater [1, 131], environment is often not flat. In such a situation, deployed sensors would form a three-dimensional wireless sensor network, which has a mathematical model, the unit ball graph. A unit ball graph consists of vertices lying in the three-dimensional Euclidean space. There exists an edge between two vertices u and v if and only if the distance between them, $d(u,v)$ is at most one.

Many approximation algorithms for MIN-CDS in unit disk graphs can be extended to unit ball graphs [127, 132], e.g., the two-stage algorithm in which, at the first stage, a maximal independent set is constructed and at the second stage, the maximal independent set is connected into a CDS. However, analysis of those approximations is a little harder than those in unit disk graphs because, instead of disk packing, spherical packing is required to study.

A set of points is said to be *independent* if every pair of points has a distance bigger than one. How many independent points can be packed into a disk of radius one? It was determined so easily in Sect. 3.2 that the answer is five. However, a similar problem in three-dimensional Euclidean space may not be so easy to answer. How many independent points can be packed into a ball of radius one? This question is closely related to the well-known problem proposed in 1694 [130].

In 1694, Isaac Newton and David Gregory discussed a kissing problem in the Cambridge University. While Newton claimed that a unit ball can touch (or kiss) at most twelve unit balls, Gregory believed that thirteen is possible. This discussion results in a well-known problem as follows.

Gregory–Newton Problem. Can a sphere touch thirteen spheres of the same size?

The answer of the Gregory–Newton problem was given by Hoppe [65] in 1874. It is proved that a sphere can touch at most 12 spheres of the same size. On the other side, there exists a sphere which can touch 12 spheres of the same size. For an example, consider an icosahedron which has 12 vertices. Let a be its edge length and r the radius of the circumscribed sphere of the icosahedron. Then $a/r = 1/\sim \frac{2\pi}{5} > 1$. This means that if place 13 spheres with radius $r/2$ at the center and vertices of the icosahedron, then the 12 spheres at vertices kiss the sphere at the center. Moreover, the 12 spheres at vertices do not touch each other.

Actually, k independent points can be packed into a ball of radius one if and only if a sphere can touch k spheres of the same size and those k spheres do not touch each other. Therefore, as a consequence of the solution of the Gregory–Newton problem, it is easy to see that the maximum number of independent points that can be packed into a ball of radius one is 12.

Kim et al. [68] extended the approach for solving the Gregory–Newton problem to two balls. They showed that if two balls with radius one have their centers within distance one, then there exist at most 20 independent points lying on their boundary surface. A consequence is that they cannot touch more than 20 balls of the same size. However, it is not easy from this result to derive a nontrivial upper bound for the number of independent points which can be packed in such two balls. Therefore, an interesting open problem is left.

Open Problem 4.1.1. *How many independent points can two balls with radius one and with center distance at most one contain?*

Since a ball with radius one can contain at most 12 independent points and the center of a ball is independent from any point not in the ball, two balls with radius one and having their centers within distance one would contain at most 23 independent points. Any upper bound less than 23 is nontrivial.

The unit disk graph and the unit ball graph are typic examples of growth-bounded graphs. Many approximation algorithms for unit disk graphs and unit ball graphs can be generalized for growth-bounded graphs [58]. Especially, PTAS exists for MIN-CDS in growth-bounded graphs [58].

4.2 Gregory–Newton Problem

The answer of the Gregory–Newton problem is presented in this section. To do so, let $\text{ball}_r(o)$ and $\text{sphere}_r(o)$ denote the ball and the sphere, respectively, with radius r and center o.

Theorem 4.2.1 (Hoppe [65]). *A sphere can touch at most twelve spheres of the same size.*

Proof. Suppose the unit sphere $\text{ball}_{0.5}(o)$ touches k unit spheres with centers $x_1, \ldots, x_k$ which are on the sphere $\text{sphere}_1(o)$. Let $\|x_i, x_j\|_s$ denote the geodesic

4.2 Gregory–Newton Problem

distance between x_i and x_j on sphere$_{0.5}(o)$. Then for any $1 \leq i < j \leq k$,

$$\|x_i, x_j\|_s \geq \pi/3.$$

Connect x_i and x_j by a geodesic arc if and only if

$$\|x_i, x_j\|_s \leq \arccos 1/7.$$

This would result in a network G on sphere$_{0.5}(o)$ with the following properties:

(P1) No two arcs cross each other.
(P2) Every node has degree at most five.

Above two properties are due to the following facts which can be proved easily by spherical trigonometry.

(F1) A spherical quadrilateral with side length at least $\pi/3$ cannot have both diagonal with length smaller than $\pi/2$.
(F2) At any node, two arc in G form an angle larger than $\pi/3$.

We can also assume that the network G is connected since all unit balls touched by ball$_{0.5}(o)$ can be put to touch each other.

The network G divides sphere$_{0.5}(o)$ into several polygons. Let A_n denote the minimum area of a polygon with n sides, that is, an n-gon. Then the A_3 is reached by an equilateral triangle and hence,

$$A_3 = 0.5512\ldots,$$

A_4 is reached by an equilateral quadrilateral with side length $\pi/3$ and one diagonal length $\arccos 1/7$ and hence,

$$A_4 = 1.3338\ldots,$$

A_5 is reached by an equilateral pentagon with side length $\pi/3$ and two coterminous diagonal length $\arccos 1/7$ and hence,

$$A_5 = 2.2261\ldots.$$

Moreover, it can be proved by induction on n that for $n \geq 5$,

$$A_n \geq 0.5725 + (n-2)A_3.$$

Let f_i be the number of i-gons. Then

$$4\pi \geq 0.5512 \cdot f_3 + 1.3338 \cdot f_4 + 2.2261 \cdot f_5 + \cdots$$
$$= 0.5512(f_3 + 2f_4 + 3f_5 + \cdots) + 0.2314 \cdot f_4 + 0.5725(f_5 + \cdots).$$

Let e be the number of arcs. By Euler's formula,

$$4k - 4 = 2e - 2(f_3 + f_4 + f_5 + \cdots)$$
$$= 3f_3 + 4f_4 + 5f_5 + \cdots - 2(f_3 + f_4 + f_5 + \cdots)$$
$$= f_3 + 2f_4 + 3f_5 + \cdots.$$

Therefore,

$$4\pi \geq 0.5512(2k - 4) + 0.2314 f_4 + 0.5725(f_5 + \cdots).$$

Thus,

$$2k - 4 \leq \frac{4\pi}{0.5512} = 22.79,$$

that is,

$$k \leq 13.$$

Assume $k = 13$. Then

$$4\pi \geq 0.5512 \times 22 + 0.2314 f_4 + 0.5525(f_5 + \cdots),$$

that is,

$$0.44 \geq 0.2314 f_4 + 0.5725(f_5 + \cdots).$$

It follows that $f_4 = 0$ or 1 and $f_5 = \cdots = 0$.

If $f_4 = 0$, then $f_3 = 2e/3$ and hence by Euler's formula, $13 + 2e/3 = e + 2$. Therefore, $e = 33$. Since $66/13 > 5$, there exists at least one node with degree six, contradicting the property (P2).

If $f_4 = 1$, then $f_3 = (2e - 4)/3$ and by Euler's formula, $13 + (1 + (2e - 4)/3) = e + 2$. Hence, $e = 32$, $f_3 = 20$. Since $64 = 12 \times 5 + 4$, G must have one node with degree 4 and 12 nodes with degree 5. Such a network does not exist. $\square$

As a corollary, we have

Corollary 4.2.2. *A ball with radius one can pack at most twelve independent points where two points are independent if and only if their Euclidean distance is bigger than one.*

Proof. Suppose $ball_1(o)$ contains k independent points $y_1, \ldots, y_k$. Let x_i be the intersection point of the half line oy_i and the sphere $sphere_1(o)$. Then by the same argument in the proof of Theorem 4.2.1, one can prove $k \leq 12$. $\square$

With this corollary, one can show the following.

Theorem 4.2.3 (Butenko and Ursulenko [9]). *For any unit ball graph G,*

$$\alpha(G) \leq 11\gamma_c(G) + 1.$$

Proof. Similar to the proof of Theorem 3.4.2. $\square$

Corollary 4.2.4. *There is a polynomial-time approximation for* MIN-CDS *in unit ball graphs which produces a CDS of size at most* $(13 + \ln 10)\gamma_c(G) + 1$ *where G is the input unit ball graph.*

Proof. It follows immediately from Theorems 3.3.2 and 4.2.3. □

4.3 Independent Points in Two Balls

Kim et al. [68] extended the approach for solving the Gregory–Newton problem to study the following problem: How many independent points can lie on boundary of the union of two balls with radius one and having their centers within distance at most one? In this section, we outline their work.

Let $B_1 = \text{ball}_1(o_1)$ and $B_2 = \text{ball}_1(o_2)$ with $d(o_1, o_2) \leq 1$. Denote by $\text{Sur}(B_1 \cup B_2)$ the boundary of $B_1 \cup B_2$, that is, $\text{Sur}(B_1 \cup B_2) = (\text{sphere}_1(o_1) \setminus B_2) \cup (sphere_1(o_2) \setminus B_1)$. Let C be the intersection circle of $\text{sphere}_1(o_1)$ and $\text{sphere}_1(o_2)$ and u the center of C. First, they showed that $\text{Sur}(B_1 \cup B_2)$ can have at most 20 independent points.

To do so, suppose $x_1, \ldots, x_k$ are independent points on $\text{Sur}(B_1 \cup B_2)$. Let $\|x_i, x_j\|_s$ denote the geodesic distance between x_i and x_j. For x_i and x_j with $\|x_i, x_j\|_s \leq \frac{3}{\pi} \arccos 1/7$, connect x_i and x_j by an arc as follows: If x_i and x_j belong to the same ball, then connect x_i and x_j by a geodesic arc. If $x_i \in B_1$ and $x_j \in B_2$, then connect x_i and x_j by a geodesic arc $x_i w$ on $\text{sphere}_1(o_1)$ and a geodesic arc $w x_j$ on $\text{sphere}_1(o_2)$ where w is an intersection point of the circle C and the plane passing through x_i, x_j and the center u of the circle C (Note: There are two intersection points, and w is the one which gives shorter total length of $x_i w$ and $w x_j$).

Above geodesic arcs would result in a network G without crossing arcs, which divide $\text{Sur}(B_1 \cup B_2)$ into polygons. Let P_n denote a polygon with n sides, that is, an n-gon. P_n is called a *regular* polygon if it lies completely in one sphere and a *striding* polygon if it has a part in $\text{sphere}_1(o_1)$ and also a part in $\text{sphere}_1(o_2)$. By routing calculation, one can obtain the following.

Lemma 4.3.1. *Let A_n be the minimum area of a regular polygon P_n. Then*

$$A_3 = 0.5512\ldots,$$
$$A_4 = 1.3338\ldots,$$
$$A_5 = 2.2261\ldots,$$
$$A_n \geq (n-2)A_3 \text{ for } n \geq 3.$$

Let $\tilde{A}_n$ be the minimum area of a striding polygon P_n. Then

$$\tilde{A}_3 = 0.4076\ldots,$$
$$\tilde{A}_4 = 0.9949\ldots,$$

$$\tilde{A}_5 = 1.8732\ldots,$$
$$\tilde{A}_n \geq (n-2)\tilde{A}_3 \quad \text{for } n \geq 3.$$

Now, it is ready to show the following.

Theorem 4.3.2 (Kim et al. [68]). $\text{Sur}(B_1 \cup B_2)$ *can contain at most 20 independent points.*

Proof. First, we move B_2 away from B_1 until $|o_1 o_2| = 1$. Note that this move would preserve the independence of points $x_1, \ldots, x_k$. However, in this position, the arc length of the circle C is $\sqrt{3}\pi$.

Let e be the number of edges in the network G. Let f be the number of polygons on $\text{Sur}(B_1 \cup B_2)$ divided by G, and f_i the number of i-gons. By Euler's formula,

$$\begin{aligned} 2(k-2) &= 2(e-f) \\ &= 3f_3 + 4f_4 + 5f_5 + \cdots - 2(f_3 + f_4 + f_5 + \cdots) \\ &= f_3 + 2f_4 + 3f_5 + \cdots. \end{aligned}$$

Let $\tilde{f}_i$ denote the number of striding i-gons. Note that $\text{Sur}(B_1 \cup B_2)$ has area at most 6π.

$$\begin{aligned} 6\pi &\geq A_3(f_3 - \tilde{f}_3) + \tilde{A}_3 \tilde{f}_3 + A_4(f_4 - \tilde{f}_4) + \tilde{A}_4 \tilde{f}_4 \\ &\quad + A_5(f_5 - \tilde{f}_5) + \tilde{A}_5 \tilde{f}_5 + \cdots \\ &\geq A_3(f_3 + 2f_4 + 3f_5 + \cdots) - (A_3 - \tilde{A}_3)(\tilde{f}_3 + 2\tilde{f}_4 + \tilde{f}_5 + \cdots) \\ &= A_3(2k-2) - (A_3 - \tilde{A}_3)(\tilde{f}_3 + 2\tilde{f}_4 + \tilde{f}_5 + \cdots). \end{aligned}$$

First, consider the case that $\tilde{f}_4 = \tilde{f}_5 = \cdots = 0$. With a geometric calculation, it is not hard to show that when the circle C passes through two adjacent striding triangles, it leaves a trace with arc length at least 0.8744. Since C has length $\sqrt{3}\pi$ and $\frac{\sqrt{3}\pi}{0.8744} = 6.223\ldots$, there are at most 12 striding triangles. Hence,

$$6\pi \geq A_3(2k-2) - (A_3 - \tilde{A}_3)12,$$

that is,

$$k \leq 20.792.$$

Therefore, $k \leq 20$.

In general, each striding n-gon for $n > 3$ can be divided into smaller polygons satisfying the following properties:

- Every i-gon for $i > 3$ is regular.
- Every striding triangle has area at least $\tilde{A}_3$.
- Every regular i-gon has area at least $(i-2)A_3$.

Note that every new edge has length longer than $\frac{3}{\pi}\arccos 1/7$. Therefore, it is still true that the circle C passing through two adjacent striding triangles leaves a piece of length at least 0.8744 inside of them. This means that this case can be reduced to the previous case. We leave the detail about the partition of i-gons for the reader to refer to the original paper [68]. □

The difficulty for using the above result to attack Open Problem 4.1.1 is as follows: Let B_1 and B_2 be two balls with radius one. Suppose their centers have distance at most one. Let I be the set of independent points contained in $B_1 \cup B_2$. Consider each point $v \in I$. If $v \in B_1 \setminus B_2$, then draw a radial from o_1 through v to intersect $\mathrm{Sur}(B_1 \cup B_2)$ at point x. If $v \in B_2 \setminus B_1$, then draw a radial from o_2 through v to intersect $\mathrm{Sur}(B_1 \cup B_2)$ at point x. Note that all x's obtained in this way are independent in B_1 and also independent in B_2. However, they may not be independent in $B_1 \cup B_2$. In other words, two x's lying in different spheres may not have distance larger than one.

4.4 Growth-Bounded Graphs

Consider a graph $G = (V, E)$ with a distance function $\mathrm{dist}(u, v)$ defined to be the number of edges on the shortest path between u and v. For any two disjoint subsets A and B of nodes, $\mathrm{dist}(A, B) = \min_{u \in A, v \in B} \mathrm{dist}(u, v)$. For any subgraph H, $\mathrm{dist}_H(u, v)$ is the number of edges in the shortest path between u and v through H and $\mathrm{dist}_H(A, B) = \min_{u \in A, v \in B} \mathrm{dist}_H(u, v)$.

For any vertex $v \in V$, define $N^r(v) = \{u \in V \mid \mathrm{dist}(u, v) \leq r\}$. For any subset V' of V, $G[V']$ denotes the subgraph induced by V' and $MaxIS(V')$ denotes the maximum independent subset in $G[V']$. A subset A of nodes is said to be *connected* if $G[A]$ is connected. For simplicity of speaking, the statement "connected components of A" may also be used instead of "the node sets of connected components of $G[A]$".

A graph $G = (V, E)$ is called a *growth-bounded graph* if there exists a polynomial function $f(\cdot)$ such that for any $v \in V$ and $r > 0$,

$$|MaxIS(N^r(v))| \leq f(r).$$

The unit disk graph and the unit ball graph are growth-bounded graphs. In fact, for the unit disk graph $G = (V, E)$,

$$|MaxIS(N^r(v))| \leq \frac{\pi(r+0.5)^2}{\pi 0.5^2} = (2r+1)^2,$$

and for the unit ball graph $G = (V, E)$,

$$|MaxIS(N^r(v))| \leq \frac{(4/3)\pi(r+0.5)^3}{(4/3)\pi 0.5^3} = (2r+1)^3.$$

A graph $G = (V,E)$ is an α-*quasi unit disk graph* for a given α, $0 < \alpha \leq 1$, if all vertices lie in the Euclidean plane and $\|u,v\| \leq \alpha$ implies $(u,v) \in E$ and $\|u,v\| > 1$ implies $(u,v) \notin E$. It was claimed that the quasi unit disk graph is more close to wireless sensor networks in the real world [70]. Note that the quasi unit disk graph is also a growth-bounded graph since

$$|MaxIS(N^r(v))| \leq \frac{\pi(r+0.5)^2}{\pi 0.5^2} \pi(0.5\alpha)^2 = (2r+1)^2/\alpha^2.$$

In this section and the next section, it is always assumed that the studied $G = (V,E)$ is connected. For such type of growth-bounded graphs, there are some important properties established by Gfeller and Vicari [58] which will be introduced in this section.

For any $A \subseteq V$, define $\mathcal{C}(A)$ to be a subset of $N(A)$ with the minimum cardinality such that for every connected component B of $G[A]$, $G[\mathcal{C}(A)]$ has a connected component dominating B. The following are two properties regarding $\mathcal{C}(A)$.

Lemma 4.4.1. *Suppose $G = (V,E)$ is a growth-bounded graph with bounding function f. Then there is a polynomial $p(r) \leq 3 \cdot f(r) - 2$ such that for any $v \in V$, $|\mathcal{C}(N^r(v))| \leq p(r)$ for any $r > 0$.*

Proof. Let M be an MIS of $N^r(A)$. Then $|M| \leq f(r)$. For each connected component B of $G[N^r(A)]$, suppose $C_1, \ldots, C_k$ are all connected components of $G[M]$ such that each C_i has a node either in B or adjacent to a node in B. Then $C_1 \cup \cdots \cup C_k$ dominates B. If $k \geq 2$, then there must exist two connected components C_i and C_j with $i \neq j$ such that $dist_{G[B]}(C_i, C_j) \leq 3$. Indeed, for contradiction, suppose such C_i and C_j do not exist. Choose a node $u \in B$ such that $dist(u, C_1) = 2$, then u is not adjacent to any C_i for $i = 1, \ldots, k$, contradicting the fact that $C_1 \cup \cdots \cup C_k$ dominates B. Now, add two nodes into M to connect C_i and C_j into one connected component. In this way, all $C_1, \ldots, C_k$ can be connected into one connected component by adding at most $2(k-1)$ nodes. Therefore, by adding at most $2(|M|-1)$ nodes, M can be modified to satisfy the property that for any connected component B of $G[A]$, $G[M]$ contains a connected component dominating B. Therefore, $|\mathcal{C}(N^r(v))| \leq 3f(r) - 2$. □

Lemma 4.4.2. *Consider a growth-bounded graph $G = (V,E)$ with bounding function f. Then for any subset A of nodes, $|MaxIS(A)| \leq f(1) \cdot |\mathcal{C}(A)|$.*

Proof. Every node in $MaxIS(A)$ is in $N(v)$ for some $v \in \mathcal{C}(A)$ and for every node $v \in \mathcal{C}(A)$, $MaxIS(N(v)) \leq f(1)$. Therefore, $|MaxIS(A)| \leq f(1) \cdot |\mathcal{C}(A)|$.

For any $V' \subseteq V$, define $\Gamma^r(v,V') = \{u \in V' \mid dist_{G[V']}(u,v) \leq r\}$. □

Lemma 4.4.3. *Let $V' \subseteq V$. Then, for any growth-bounded graphs $G = (V,E)$ with a bounding function f,*

$$|MaxIS(N(\Gamma^r(v,V')) \setminus \Gamma^r(v,V'))| \leq f(2)|MaxIS(\Gamma^r(v,V') \setminus \Gamma^{r-1}(v,V'))|$$

for any $r \geq 1$ and $v \in V$.

4.4 Growth-Bounded Graphs

Proof. Each node u in $MaxIS(N(\Gamma^r(v,V')) \setminus \Gamma^r(v,V'))$ has a neighbor in $\Gamma^r(v,V') \setminus \Gamma^{r-1}(v,V')$ which is also a neighbor of some node w in $MaxIS(\Gamma^r(v,V') \setminus \Gamma^{r-1}(v,V'))$. Therefore, $u \in N^2(w)$. Since $N^2(w)$ can contain at most $f(2)$ nodes in $MaxIS(N(\Gamma^r(v,V')) \setminus \Gamma^r(v,V'))$, the inequality holds. □

Lemma 4.4.4. *For any class of growth-bounded graphs with the same bounding function f and for any $\varepsilon > 0$, there exists a constant $R_f(\varepsilon) = O(1/\varepsilon \cdot \log(1/\varepsilon))$ such that for every graph $G = (V,E)$ in this class, every subset $V' \subseteq V$ and every vertex $v \in V$,*

$$|MaxIS(N(\Gamma^r(v,V')) \setminus \Gamma^r(v,V'))| \le \varepsilon \cdot |MaxIS(\Gamma^r(v,V'))|$$

for some $r \le R_f(\varepsilon)$.

Proof. For contradiction, suppose such a $R_f(\varepsilon)$ does not exists for some f and some ε. Then for arbitrarily large r, there exists a graph $G_r = (V_r, E_r)$ and a subset $V'_r \subseteq V_r$ in the class such that

$$|MaxIS(N(\Gamma^{r'}(v,V'_r)) \setminus \Gamma^{r'}(v,V'_r))| > \varepsilon \cdot |MaxIS(\Gamma^{r'}(v,V'_r))|$$

for all $0 \le r' \le r$. By Lemma 4.4.3,

$$|MaxIS(\Gamma^{r'}(v,V'_r) \setminus \Gamma^{r'-1}(v,V'_r))| > \bar{\varepsilon} \cdot |MaxIS(\Gamma^{r'}(v,V'_r))|$$
$$\ge \bar{\varepsilon} \cdot |MaxIS(\Gamma^{r'-2}(v,V'_r))|$$

for $\bar{\varepsilon} = \varepsilon/f(2)$. Thus, for $2 \le r' \le r$,

$$|MaxIS(\Gamma^{r'}(v,V'_r))|$$
$$\ge |MaxIS(\Gamma^{r'}(v,V'_r) \setminus \Gamma^{r'-1}(v,V'_r))| + |MaxIS(\Gamma^{r'-2}(v,V'_r))|$$
$$> (1+\bar{\varepsilon}) \cdot |MaxIS(\Gamma^{r'-2}(v,V'_r))|$$
$$> (1+\bar{\varepsilon})^2 \cdot |MaxIS(\Gamma^{r'-4}(v,V'_r))|$$
$$> \cdots$$
$$> (1+\bar{\varepsilon})^{\lfloor r'/2 \rfloor}.$$

Note that $|MaxIS(\Gamma^{r'}(v,V'_r))| \le f(r')$. Therefore,

$$f(r) \ge (1+\bar{\varepsilon})^{\lfloor r/2 \rfloor}$$

for all $r \ge 2$, which is impossible since $f(r)$ is a polynomial.

From above proof, it is easy to see that if R satisfies

$$f(R) < (1+\bar{\varepsilon})^{\lfloor r/2 \rfloor} \quad (4.1)$$

then there must exist $0 < r \leq R$ such that

$$|MaxIS(N(\Gamma^r(v,V'))\setminus \Gamma^r(v,V'))| \leq \varepsilon \cdot |MaxIS(\Gamma^r(v,V'))|.$$

Suppose $f(R) \leq R^\alpha$. Note that $(1+\varepsilon)^{\frac{1}{1+1/\varepsilon}} > e$. Therefore, to make (4.1) hold, it suffices to have

$$R^\alpha < e^{\frac{\varepsilon}{1+\varepsilon}\cdot(\frac{R-1}{2})}.$$

Choose $R = 1 + 4(1+2\alpha)(1/\varepsilon)\ln(1/\varepsilon)$. Then for $\varepsilon < \frac{1}{1+4(1+2\alpha)}$,

$$R^\alpha < (1/\varepsilon)^{1+2\alpha} < (1/\varepsilon)^{(1+2\alpha)\cdot \frac{2}{1+\varepsilon}} = e^{\frac{\varepsilon}{1+\varepsilon}\cdot \frac{R-1}{2}}. \qquad \square$$

Lemma 4.4.5. *Suppose that*

$$|MaxIS(N(\Gamma^r(v,V'))\setminus \Gamma^r(v,V'))| \leq \varepsilon \cdot |MaxIS(\Gamma^r(v,V'))|$$

holds for some $\varepsilon > 0$ and some $r > 0$. Then for $\varepsilon' = \varepsilon 4 f(3)f(1)$,

$$|\mathcal{C}(\Gamma^{r+4}(v,V'))| \leq (1+\varepsilon')|\mathcal{C}(\Gamma^r(v,V'))|.$$

Proof. Each $u \in MaxIS(\Gamma^{r+4}(v,V')) \setminus \Gamma^r(v,V')$ is within distance 3 from a node $w \in N(\Gamma^r(v,V')) \setminus \Gamma^r(v,V')$. Therefore,

$$|MaxIS(\Gamma^{r+4}(v,V')) \setminus \Gamma^r(v,V')|$$
$$\leq f(3)|MaxIS(N(\Gamma^r(v,V'))\setminus \Gamma^r(v,V'))|$$
$$\leq \varepsilon f(3) \cdot |MaxIS(\Gamma^r(v,V'))|.$$

From the proof of Lemma 4.4.1, it can be seen that

$$|\mathcal{C}(\Gamma^{r+4}(v,V')\setminus \Gamma^r(v,V'))| \leq 3\varepsilon f(3) \cdot |MaxIS(\Gamma^r(v,V'))|.$$

Note that the number of connected components of $\mathcal{C}(\Gamma^{r+4}(v,V')\setminus \Gamma^r(v,V'))$ is at most

$$|MaxIS(\Gamma^{r+4}(v,V'))\setminus \Gamma^r(v,V')| \leq \varepsilon f(3) \cdot |MaxIS(\Gamma^r(v,V'))|.$$

Thus, it needs to add at most $\varepsilon f(3)|MaxIS(\Gamma^r(v,V'))|$ nodes to connect $\mathcal{C}(\Gamma^r(v,V'))$ and $\mathcal{C}(\Gamma^{r+4}(v,V')\setminus \Gamma^r(v,V'))$ together into A such that for any connected component B of $\Gamma^{r+4}(v,V')$, $G[A]$ has a connected component dominating B. Therefore,

$$|\mathcal{C}(\Gamma^{r+4}(v,V'))|$$
$$\leq |\mathcal{C}(\Gamma^r(v,V'))| + 4\varepsilon f(3)|MaxIS(\Gamma^r(v,V'))|$$
$$\leq (1+\varepsilon')|\mathcal{C}(\Gamma^r(v,V'))|. \qquad \square$$

4.5 PTAS in Growth-Bounded Graphs

Gfeller and Vicari [58] designed a PTAS for MIN-CDS in growth-bounded graphs as follows.

Algorithm GBG.
input a growth-bounded graph $G = (V,E)$ and a number $\varepsilon > 0$;
$D \leftarrow \emptyset$;
$V' \leftarrow V$;
while $V \neq \emptyset$ **do begin**
    $r \leftarrow 0$;
    while $|MaxIS(N(\Gamma^r(v,V'))\setminus \Gamma^r(v,V'))| > \varepsilon|MaxIS(\Gamma^r(v,V'))|$ **do**
        $r \leftarrow r+1$;
    $D \leftarrow D \cup \mathcal{C}(\Gamma^{r+4}(v,V'))$;
    $V' \leftarrow V' \setminus \Gamma^{r+2}(v,V')$;
end-while;
output D.

The following is the analysis of this algorithm.

Lemma 4.5.1. *Algorithm GBG outputs a CDS D.*

Proof. Clearly D is a dominating set. In the following, the connectivity of $D = \cup \mathcal{C}(T_i)$ is proved. First, prove a claim that for $u,u' \in D$ with $\text{dist}(u,u') = 2$, u and u' are connected in $G[D]$. Since $\text{dist}(u,u') = 2$, there exists a node w such that u and u' are in $N(w)$. Suppose that among nodes in $N(w)$, s is the first one deleted from V' in Algorithm GBG. Then when $s \in \Gamma^{r+2}(v,V')$, $N(w) \subseteq \Gamma^{r+2}(v,V')$. Therefore, u and u' are in the same connected component of $G[\mathcal{C}(\Gamma^{r+4}(v,V'))]$ and hence in the same connected component of $G[D]$. By Lemma 4.4.1, they are connected by a path in D with length at most $3f(r+4) - 3 \leq 3(R_f(\varepsilon) + 4) - 3$.

For $u,u' \in D$ with $\text{dist}(u,u') = 3$, suppose (u,w,w',u') is the path connecting u and u' in G. Suppose w is deleted from V' no later than w', that is, when $w \in \Gamma^r(v,V')$, $w' \in \Gamma^{r+4}(v,V')$. Therefore, w and w' are dominated by a connected component of $G[\mathcal{C}(\Gamma^{r+4}(v,V'))]$. Hence, there is a path $(w,y_1,\ldots,y_k,w')$ of length at most $3f(r+4) - 1 (\leq 3f(R_f(\varepsilon) + 4) - 1)$, connecting w and w' through D. By the above claim, u and y_1 are connected by a path in D with length at most $3f(R_f(\varepsilon) + 4) - 3$ and y_k and u' are connected by a path in D with length at most $3f(R_f(\varepsilon) + 4) - 3$. Therefore, u and u' are connected by a path in D with length at most $9f(R_f(\varepsilon) + 4) - 9$.

Now, note that if $G[D]$ is not connected, then there exist two connected components within distance three since D is a dominating set. However, it has been proved that for any two nodes $u,u' \in D$ with $\text{dist}(u,v) \leq 3$, u and u' are connected in D. Therefore, $G[D]$ has to be connected. □

Lemma 4.5.2. *Let S_i be the set $\Gamma^r(v,V')$ in the ith iteration of the outer while-loop of Algorithm GBG. Suppose the outer while-loop runs totally k times. Then*

$$\sum_{i=1}^{k}|\mathcal{C}(S_i)| \leq (1+2\varepsilon f(1))|\mathcal{C}(V)|.$$

Proof. Note that all $N(S_1),\ldots,N(S_k)$ are disjoint. Thus, it is sufficient to show $|\mathcal{C}(S_i)| \leq |\mathcal{C}(V) \cap N(S_i)|$. To do so, it suffices to modify the connectivity of $\mathcal{C}(V) \cap N(S_i)$ and estimate how many nodes required to do the modification, since $\mathcal{C}(V) \cap N(S_i)$ dominates S_i.

Suppose $\mathcal{C}(V) \cap N(S_i)$ has z connected component. Each connected component must have a node lying in $N(S_i) \setminus S_i$ in order to establish the global connectivity of $\mathcal{C}(V)$. Therefore,

$$z \leq |MaxIS(N(S_i) \setminus S_i)| \leq \varepsilon \cdot |MaxIS(S_i)|.$$

This means that it is sufficient to add at most $2\varepsilon \cdot |MaxIS(S_i)|$ nodes into $\mathcal{C}(V) \cap N(S_i)$ in order to make it satisfy the property that for every connected component B of S_i, there is a connected component of modified $\mathcal{C}(V) \cap N(S_i)$ that dominates B. This means that

$$|\mathcal{C}(S_i)| \leq |\mathcal{C}(V) \cap N(S_i)| + 2\varepsilon |MaxIS(S_i)|.$$

Moreover, by Lemma 4.4.1,

$$|MaxIS(S_i)| \leq f(1)|\mathcal{C}(V) \cap N(S_i)|.$$

Hence,

$$|\mathcal{C}(S_i)| \leq (1+2\varepsilon f(1))|\mathcal{C}(V) \cap N(S_i)|. \qquad \square$$

Lemma 4.5.3. *Algorithm GBG runs in time* $n^{O(f(R_f(\varepsilon)))}$.

Proof. The inner while-loop runs at most $R_f(\varepsilon)$ times. In each iteration of the inner while-loop, computing $MaxIS(N(\Gamma^r(v,V')) \setminus \Gamma^r(v,V'))$ and $MaxIS(\Gamma^r(v,V'))$ takes $n^{O(f(R_f(\varepsilon)))}$ time. Therefore, Algorithm GBG runs in $O(R_f(\varepsilon) \cdot n^{O(f(R_f(\varepsilon)))}) = n^{O(f(R_f(\varepsilon)))}$ time. $\qquad \square$

Lemma 4.5.4. *The output D of Algorithm GBG is a* $(1+O(\varepsilon))$-*approximation for* MIN-CDS *in growth-bounded graphs.*

Proof. Suppose the outer while-loop of Algorithm GBG runs k times. Denote by S_i and T_i, respectively, the set $\Gamma^r(v,V')$ and the set $\Gamma^{r+4}(v,V')$ in the ith iteration of the outer while-loop of Algorithm GBG. By Lemma 4.4.5,

$$|\mathcal{C}(T_i)| \leq (1+\varepsilon')|\mathcal{C}(S_i)|$$

for $1 \leq i \leq k$, and by Lemma 4.5.2,

$$\sum_{i=1}^{k}|\mathcal{C}(S_i)| \leq (1+2\varepsilon f(1))|\mathcal{C}(V)|.$$

4.5 PTAS in Growth-Bounded Graphs

Therefore,

$$\begin{aligned}
|D| &= |\cup_{i=1}^{k} \mathcal{C}(T_i)| \\
&\leq \sum_{i=1}^{k} |\mathcal{C}(T_i)| \\
&\leq (1+\varepsilon') \sum_{i=1}^{k} |\mathcal{C}(S_i)| \\
&\leq (1+\varepsilon')(1+2\varepsilon f(1))|\mathcal{C}(V)|.
\end{aligned}$$

□

Theorem 4.5.5 (Gfeller and Vicari [58]). *Algorithm GBG is a PTAS for* MIN-CDS *in growth-bounded graphs.*

Proof. It follows immediately from Lemmas 4.5.3 and 4.5.4. □

Chapter 5
Weighted CDS in Unit Disk Graph

I think the weight really got the best of him today.
JODY PETTY

5.1 Motivation and Overview

It was open for many years whether MINW-CDS in unit disk graphs has a polynomial-time constant-approximation or not. Ambühl et al. [2] discovered the first one. Their solution consists of two stages. At the first stage, they construct a dominating set which is a 72-approximation for the minimum-weight dominating set problem in unit disk graphs as follows.

MINW-DS in Unit Disk Graphs: Given a unit disk graph $G = (V, E)$ with vertex weight $w : V \to R^+$, find a dominating set with minimum total weight.

In the second stage, they connect the dominating set into a CDS with additional cost $12opt_{WCDS}$ where opt_{WCDS} is the minimum weight of a CDS. Putting together, they obtained a polynomial-time 94-approximation for MINW-CDS in unit disk graphs.

Huang et al. [66] discovered a new technique on partition, called *double partition*. With the new technique, they obtained a polynomial-time $(6+\varepsilon)$-approximation for MINW-DS in unit disk graphs. Later, the approximation for MINW-DS in unit disk graphs received further improvements, from performance ratio $6+\varepsilon$ to $5+\varepsilon$ by Dai and Yu [27], to $4+\varepsilon$ by Zou et al. [134] and independently by [46], and to 3.63 by Willson et al. [114].

Connecting a weighted dominating set into a weighted CDS is equivalent to solving NODE-WEIGHTED STEINER TREE in unit disk graphs.

In general graphs, it is unlikely for NODE-WEIGHTED STEINER TREE to have a polynomial-time constant-approximation [69]. However, in unit disk graphs, the situation is different. Actually, the work of Ambühl et al. [2] means that there is a polynomial-time 12-approximation for NODE-WEIGHTED STEINER TREE in unit disk graphs. Huang et al. [66] gave a polynomial-time 4-approximation. Zou et al. [133] constructed a polynomial-time 2.5ρ-approximation for NODE-WEIGHTED STEINER TREE in unit disk graphs provided that there exists a polynomial-time ρ-approximation for the minimum network Steiner tree problem. Recently, the minimum network Steiner tree problem has been found to have a polynomial-time 1.39-approximation [10]. Therefore, the approximation of Zou et al. [133] has performance 3.475. Hence, there exists a polynomial-time 7.105-approximation for MINW-CDS in unit disk graphs.

The following are still open:

Open Problem 5.1.1. *Does* MINW-CDS *in unit disk graphs have a PTAS?*

Open Problem 5.1.2. *Does* MINW-CDS *in unit ball graphs have a polynomial-time constant-approximation?*

5.2 Node-Weighted Steiner Tree

In this section, we introduce the approximation algorithm of Zou et al. [10] for NODE-WEIGHTED STEINER TREE in unit disk graphs.

Their design is motivated from the following property of optimal solutions for NODE-WEIGHTED STEINER TREE in unit disk graphs.

Lemma 5.2.1. *In a unit disk graph, for any set of terminals, there exists an optimal solution T for* NODE-WEIGHTED STEINER TREE *such that every node has degree at most five.*

Proof. Among all optimal trees for NODE-WEIGHTED STEINER TREE, we consider the one with the shortest Euclidean edge length, called the shortest optimal tree. First, note that the shortest optimal tree must have the following properties:

(a1) No two edges cross each other.
(a2) Two edges meet at a node with an angle of at least $60°$.
(a3) If two edges meet with an angle of exactly $60°$, then they have the same length.

Indeed, if anyone of the above three conditions does not hold, then we can easily find another optimal tree with shorter length.

Now, consider a shortest optimal tree T. By (a2), every node has degree at most six. Suppose T has a node u with degree exactly six, that is, u has six neighbors $v_1, v_2, \ldots, v_6$. By (a2), $\angle v_1 u v_2 = \angle v_2 u v_3 = \cdots = \angle v_6 u v_1 = 60°$. By (a3), $|uv_1| = |uv_2| = \cdots = |uv_6|$. Moreover, v_2 must have degree at most four since replacing (u,v_1) and (u,v_3) by (v_1,v_2) and (v_2,v_3), the result should still be a shortest optimal tree. Now, replace (u,v_1) by (v_1,v_2) and do similar replacement at all nodes with degree six. Then one would obtain a shortest optimal tree with node degree at most five. □

5.2 Node-Weighted Steiner Tree

Assign each edge (u,v) with the following weight:

$$w(u,v) = \frac{1}{2}(\chi_P(u)c(u) + \chi_P(v)c(v)),$$

where

$$\chi_P(u) = \begin{cases} 1, & \text{if } u \in P, \\ 0, & \text{otherwise.} \end{cases}$$

Let T^*_{node} be the optimal solution for NODE-WEIGHTED STEINER TREE in unit disk graph G, with the property that every node has degree at most five. Then

$$w(T^*_{\text{node}}) \leq 2.5 c(T^*_{\text{node}}).$$

Let T^*_{edge} be the minimum edge-weight Steiner tree on terminal set P. Then

$$w(T^*_{\text{edge}}) \leq w(T^*_{\text{node}})$$

and from [10], one can compute a 1.39-approximation T for T^*_{edge} in polynomial-time. Therefore

$$w(T) \leq 1.39 \cdot w(T^*_{\text{edge}}) \leq 3.475 \cdot c(T^*_{\text{node}}).$$

Moreover,

$$c(T) \leq w(T)$$

since each Steiner node has degree at least two in T. Therefore,

$$c(T) \leq 3.475 \cdot c(T^*_{\text{node}}).$$

Above analysis suggests the following approximation algorithm for NODE-WEIGHTED STEINER TREE for unit disk graphs.

3.475-Approximation
input: unit disk graph $G = (V,E)$ with node weight $c : V \to R^+$
and a terminal set $P \subseteq V$.
compute 1.39-approximation T for the minimum edge-weight Steiner tree on terminal set P in graph G with edge weight
$w(u,v) = \frac{1}{2}(\chi_P(u)c(u) + \chi_P(v)c(v))$ for $(u,v) \in E$;
output T.

Theorem 5.2.2 (Zou et al. [133]). *There exists a polynomial-time 3.475-approximation for* NODE-WEIGHTED STEINER TREE *in unit disk graphs.*

By this theorem, if there exists a polynomial-time τ-approximation for MINW-DS in unit disk graphs, then there exists a polynomial-time $(\tau + 3.475)$-approximation for MINW-CDS in unit disk graphs. The remaining part of this chapter will be contributed to the study of MINW-DS in unit disk graphs.

5.3 Double Partition

The partition is a classical technique to design approximation algorithms [36]. In Sect. 3.2, this technique has been used to design a PTAS for MIN CDS in unit disk graphs. From there, one may see that the approximation performance ratio and the running time have a trade-off. Indeed, the running time of $(1+\varepsilon)$-approximation is $n^{O(1/\varepsilon^2)}$. As the approximation performance ratio $1+\varepsilon$ approaches to 1, the running time $n^{O(1/\varepsilon^2)}$ increases rapidly. Meanwhile, the size of cells, $O(1/\varepsilon)$ in the partition would also increase linearly. Indeed, the design of PTAS is based on the fact that MIN-CDS is polynomial time solvable within any constant-size cell.

When applying the partition to MINW-DS, the trouble one meets is that even for a small constant-size cell, no polynomial-time algorithm for the optimal solution has been obtained so far. Only within a square of edge size at most $\sqrt{2}/2$, a polynomial-time 2-approximation exists. In such a case, the double partition technique can be employed to overcome the trouble.

Initially, the input unit disk graph is put in a square. In the first partition, the square is partitioned into blocks; each block is a small square with edge length $m\sqrt{2}/2$. In the second partition, each block is partitioned into smaller cells with edge length $\sqrt{2}/2$. The advantage of double partition is on the second one. When m is fixed, each block can be seen to contain a constant-number m^2 of cells so that many types of combinations about cells can be enumerated in polynomial-time. In this section, the first partition is introduced under the assumption that there is a ρ-approximation with running time $n^{O(m^2)}$ for the following problem.

MINW-DS on a Block B: Given a unit disk graph $G=(V,E)$ with a nonnegative node weight $w: V \to R^+$, find the minimum-weight node subset to dominate all nodes lying inside of B.

With the first partition, one shows the following.

Theorem 5.3.1 (Huang et al. [66]). *Suppose there exists a ρ-approximation for* MINW-DS *on a fixed block B with running time $n^{O(m^2)}$. Then for any $\varepsilon > 0$, there exists $(\rho + \varepsilon)$-approximation with computation time $n^{O(1/\varepsilon^2)}$ for* MINW-DS *in unit disk graphs.*

Proof. Choose $m = 12\max(1, \lceil 1/\varepsilon \rceil)$. Put input unit disk graph G into a grid with each block being an $m\mu \times m\mu$ square (Fig. 5.1). All blocks are disjoint. To do so, each block has boundary open on the left and on the top, but close on the right and on the bottom.

For each nonempty block B, compute ρ-approximation for MINW-DS on block B. Unit those ρ-approximation solutions for all nonempty block and denote this union by $A(P)$ for the partition P induced by this grid.

Now, shaft this grid in diagonal direction with distance 4 in each time. This results in $m/4$ partitions $P_1, \ldots, P_{m/4}$. Choose $A = A(P_i)$ to be the one with the minimum weight among $A(P_1), \ldots, A(P_{m/4})$. Now, one claims that $c(A(P_i)) \leq (\rho + \varepsilon)\text{opt}_{\text{WDS}}$ where opt_{WDS} is the total weight of optimal solution for MINW-DS.

5.3 Double Partition

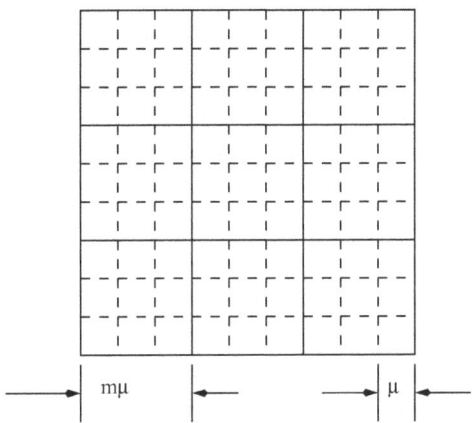

Fig. 5.1 Double partition ($\mu = \sqrt{2}/2$)

Suppose Opt_{WDS} is an optimal solution for MINW-DS in unit disk graph G. For each $v \in \text{Opt}_{\text{WDS}}$, the disk $\text{disk}_1(v)$ may intersect more than one blocks of P_i. Let $\zeta_i(v)$ be the number of blocks in partition P_i, intersecting $\text{disk}_1(v)$. Let $O(B) = \{v \in \text{Opt}_{\text{WDS}} \mid \text{disk}_1(v) \cap B \neq \emptyset\}$. Then $O(B)$ is a feasible solution for MINW-DS on block B. Therefore,

$$c(A(P_i)) \leq \sum_{B \in P_i} c(O(B)) = \text{opt}_{\text{WDS}} + \sum_{v \in \text{Opt}_{\text{WDS}}} (\zeta_i(v) - 1) c(v).$$

Note that $\text{disk}_1(v)$ can intersect at most one horizontal cutline and at most one vertical cutline of P_i. Therefore, $\zeta_i(v)$ has only three possible values. $\zeta_i(v) = 1$ if $\text{disk}_1(v)$ does not intersect any cutline of P_i, $\zeta_i(v) = 2$ if $\text{disk}_1(v)$ intersects exactly one cutline of P_i, and $\zeta_i(v) = 4$ if $\text{disk}_1(v)$ intersects two cutlines of P_i, one horizontal cutline and one vertical cutline.

Moreover, two vertical cutlines, possibly from two different partitions, have distance $2\sqrt{2} > 2$. Thus, over all partitions, each disk $\text{disk}_1(v)$ for $v \in \text{Opt}_{\text{WDS}}$ can intersect at most one vertical cutline and similarly at most one horizontal cutline. This means that for every $v \in \text{Opt}_{\text{WDS}}$,

$$\sum_{i=1}^{m/4} (\zeta_i(v) - 1) \leq 3.$$

Therefore,

$$c(A) = \min_{1 \leq i \leq m/4} c(A(P_i))$$

$$\leq \frac{1}{m/4} \sum_{i=1}^{m/4} \sum_{B \in P_i} c(O(B))$$

$$= \text{opt}_{\text{WDS}} + \frac{4}{m} \sum_{v \in \text{Opt}_{\text{WDS}}} \sum_{i=1}^{m/4} (\zeta_i(v) - 1) c(v)$$

$$\leq \text{opt}_{\text{WDS}} + \frac{12}{m} \text{opt}_{\text{WDS}}$$

$$\leq (1 + \varepsilon) \text{opt}_{\text{WDS}}. \qquad \square$$

5.4 Cell Decomposition

In the second partition, each block is partitioned into $\frac{\sqrt{2}}{2} \times \frac{\sqrt{2}}{2}$ cells also by a grid. To have all cells nonoverlapping, assume that each cell has open boundary on the right and on the top, and close on the left and on the bottom. This section is contributed to study the following problem.

MINW-DS on a Cell e: Given a unit disk graph $G = (V, E)$ with a nonnegative node weight $w : V \to R^+$, find the minimum-weight node subset to dominate all nodes lying inside of e.

The main duty of this section is to prove the following result.

Lemma 5.4.1. *There is a polynomial-time 2-approximation for* MINW-DS *in a cell e.*

The proof of this lemma is based on a decomposition of nodes in the cell e into two parts which form two polynomial-time solvable subproblems. This decomposition stems from the property of optimal solution for MINW-DS in the cell e.

Suppose $\text{Opt}(e)$ is an optimal solution for MINW-DS in a cell e. If $\text{Opt}(e)$ contains a node v lying in e, then $\text{Opt}(e) = \{v\}$ and $c(v) = \min_{u \in e} c(u)$ because any node in e is able to dominate every point of e. The difficult part of characterizing $\text{Opt}(e)$ is in the case that $\text{Opt}(e)$ does not contain any node in e.

To deal with this case, let A, B, C, D be four vertices of e and divide outside of e into eight areas NE (northeastern), NC (north-central), NW (northwestern), ME (middle-east), MW (middle-west), SE (southeastern), SC (south-central), and SW (southwestern) as shown in Fig. 5.2.

Let $V(e)$ be the set of nodes lying in the cell e. $V(e)$ will be decomposed into two parts $V(e) = V_1 \cup V_2$ ($V_1 \cap V_2 = \emptyset$) such that all points in V_1 can be dominated by nodes in $\text{Opt}_1(e) = \text{Opt}(e) \cap (N \cup S)$ where $N = NE \cup NC \cup NW$ and $S = SE \cup SC \cup SW$, and V_2 can be dominated by nodes in $\text{Opt}_2(e) = \text{Opt}(e) \cap (E \cup W)$ where $E = NE \cup ME \cup SE$ and $W = NW \cup MW \cup SW$.

Next, the existence of such a partition of $V(e)$ for $\text{Opt}(e)$ would be proved through presentation of two lemmas.

For any vertex $p \in V(e)$, let $\angle p$ be a right angle at p such that two edges intersect horizontal line AB each at an angle of $\pi/4$. Let $\Delta_{\text{south}}(p)$ denote the part of e lying inside of $\angle p$. Similarly, we can define $\Delta_{\text{north}}(p)$, $\Delta_{\text{east}}(p)$ and $\Delta_{\text{west}}(p)$ as shown in Fig. 5.3.

5.4 Cell Decomposition

Fig. 5.2 Outside of e is divided into eight areas

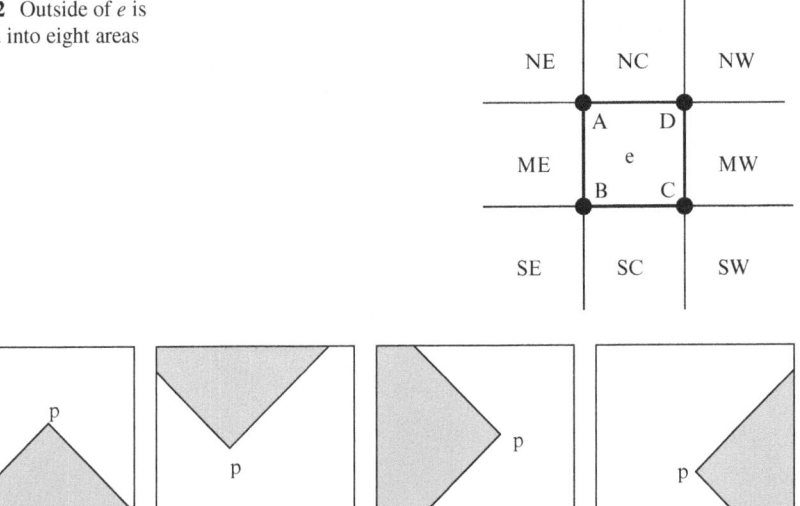

Fig. 5.3 $\Delta_{\text{south}}(p)$, $\Delta_{\text{north}}(p)$, $\Delta_{\text{east}}(p)$ and $\Delta_{\text{west}}(p)$

Lemma 5.4.2. *If p is dominated by a vertex u in area SC then every point in $\Delta_{\text{south}}(p)$ can be dominated by u. The similar statement holds for ME and $\Delta_{\text{east}}(p)$, MW and $\Delta_{\text{west}}(p)$, and NC and $\Delta_{\text{north}}(p)$.*

Proof. Note that $\Delta_{\text{south}}(p)$ is a convex polygon. It is sufficient to show that the distance from u to every vertex of $\Delta_{\text{south}}(p)$ is at most one.

Suppose v is a vertex of $\Delta_{\text{south}}(p)$ on BC (Fig. 5.4). Draw a line L' perpendicular to pv and equally divide pv. If u is below L', then $d(u,v) \leq d(u,p) \leq 1$. If u is above L', then $\angle uvp < \pi/2$ and hence $\angle uvC < 3\pi/4$. Hence, $d(u,v) < \mu/\cos\pi/4 = 1$.

A similar argument can be applied in the case that the vertex v of $\Delta_{\text{south}}(p)$ is on DA or on AB. $\square$

Consider two nodes $p, p' \in V(e)$. Suppose p is on the left of p'. Extend the left edge of $\angle p$ and the right edge of $\angle p'$ to intersect at point p''. Define $\Delta_{\text{south}}(p,p')$ to be the part of e lying inside of $\angle p''$ (Fig. 5.5). Similarly, we can define $\Delta_{\text{north}}(p,p')$.

Lemma 5.4.3. *Let K be a set of nodes which dominates $V(e)$. Suppose $p, p' \in V(e)$ are dominated by some nodes in $K \cap SC$, but neither p nor p' is dominated by any node in $K \cap (ME \cup MW)$. Then every node in $\Delta_{\text{south}}(p,p')$ can be dominated by node in $K \cap (N \cup S)$ where $N = NE \cup NC \cup NW$ and $S = SE \cup SC \cup SW$.*

Proof. By Lemma 5.4.2, it suffices to consider a node u lying in $\Delta_{\text{south}}(p,p') \setminus (\Delta_{\text{south}}(p) \cup \Delta_{\text{south}}(p'))$. For contradiction, suppose u is dominated by a node v in $K \cap (ME \cup MW)$. If $v \in ME$, then $\Delta_{\text{east}}(v)$ contains p and by Lemma 5.4.2, p is dominated by v, a contradiction. A similar contradiction can result from $v \in MW$. $\square$

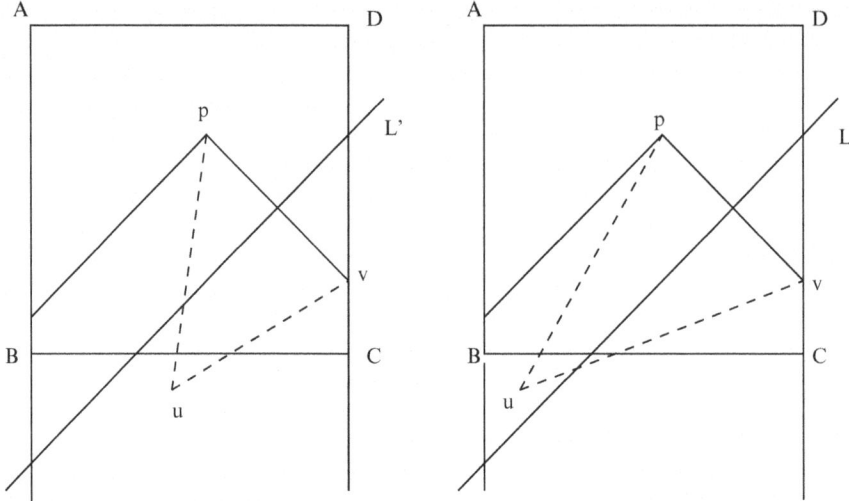

Fig. 5.4 The proof of Lemma 5.4.2

Fig. 5.5 $\Delta_{\text{south}}(p, p')$

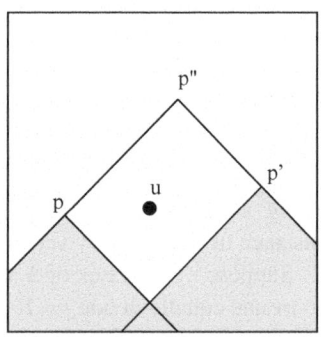

Now, it is ready to give a property of Opt(e) in case that Opt(e) $\cap V(e) = \emptyset$.

Lemma 5.4.4. *Let* Opt(e) *be an optimal solution for* MINW-DS *in the cell* e. *Suppose* Opt(e) $\cap V(e) = \emptyset$. *Then there exist four nodes* $p, p', q, q' \in V(e)$ *such that* $V_1(e) = V(e) \cap (\Delta_{\text{south}}(p, p') \cup \Delta_{\text{north}}(q, q'))$ *is dominated by* Opt$_1(e)$ = Opt(e) $\cap$ ($N \cup S$) *and* $V_2(e) = V(e) - V_1(e)$ *is dominated by* Opt$_2(e)$ = Opt(e) $\cap (E \cup W)$.

Proof. Let V_S be the set of nodes in $V(e)$, each of which can be dominated by a node in SC but not dominated by any node in $ME \cup MW$. Let p be the node in $V(S)$ such that the left edge of $\Delta(p)$ is on the leftmost position among all left edges of $\Delta(v)$ for $v \in V_S$. Let p' be the node in V_S such that the right edge of $\Delta(p')$ is on the rightmost position among all right edges of $\Delta_{\text{south}}(v)$ for $v \in V_S$. Clearly, $\Delta_{\text{south}}(p, p')$ has the following properties:

(p1) Every node in $\Delta_{\text{south}}(p, p')$ can be dominated by Opt$_1(e)$.
(p2) $V_S \subset \Delta_{\text{south}}(p, p')$.

5.4 Cell Decomposition

Similarly, let V_N be the set of nodes in $V(e)$, each of which can be dominated by a node in SC but not dominated by any node in $ME \cup MW$. One can find nodes $q, q' \in V_N$ to meet the following requirement.

(q1) Every node in $\Delta_{\text{north}}(q, q')$ can be dominated by $\text{Opt}_1(e)$.
(q2) $V_N \subset \Delta_{\text{north}}(p, p')$.

It follows from (p1) and (q1) that $V_1(e)$ is dominated by $\text{Opt}_1(e)$. If follows from (p2) and (q2) that $V_2(e)$ is dominated by $\text{Opt}_2(e)$. □

Based on Lemma 5.4.4, one can design a 2-approximation for MINW-DS in the cell e as follows.

2-Approximation for MINW-DS in a cell e
    input a weighted unit disk graph G and a cell e.
    $u \leftarrow \text{argmin}_{v \in V(e)} c(v)$;
    $V^+(e) \leftarrow \{v \in V \mid \text{disk}_1(v) \cap e \neq \emptyset\}$;
    $V_1^+ \leftarrow V^+(e) \cap (N \cup S)$;
    $V_2^+ \leftarrow V^+(e) \cap (E \cup W)$;
    $A \leftarrow \{u\}$;
    for every $\{p, p', q, q'\} \subset V(e)$
    do begin
        $V_1 \leftarrow V(e) \cap (\Delta_{\text{south}}(p, p') \cup \Delta_{\text{north}}(q, q'))$;
        $V_2 \leftarrow V(e) \setminus V_1$;
        find the minimum-weight subset O_1 of V_1^+, dominating V_1;
        find the minimum weight subset O_2 of V_2^+, dominating V_2;
        if $c(A) > c(O_1 \cup O_2)$
            $A \leftarrow O_1 \cup O_2$;
    end-for;
    output $A(e) = A$.

Clearly, if $\text{Opt}(e) \cap V(e) \neq \emptyset$, then $c(A) = c(\text{Opt}(e))$. If $\text{Opt}(e) \cap V(e) = \emptyset$, then for $\{p, p', q, q'\}$ in Lemma 5.4.4, one has

$$c(O_1) \leq c(\text{Opt}_1(e)), c(O_2) \leq c(\text{Opt}_2(e)).$$

Therefore,

$$c(A(e)) \leq c(O_1 \cup O_2) \leq c(\text{Opt}_1(e)) + c(\text{Opt}_2(e)) \leq 2c(\text{Opt}(e)).$$

The next section will show that O_1 and O_2 can be computed in polynomial-time. Therefore, the following holds.

Lemma 5.4.5. *There is a polynomial-time 2-approximation for* MINW-DS *in the cell e.*

The following result can be easily obtained based on Lemma 5.4.5.

Fig. 5.6 For two cells at ends of a diagonal, at most one has its interior intersecting a disk of radius one

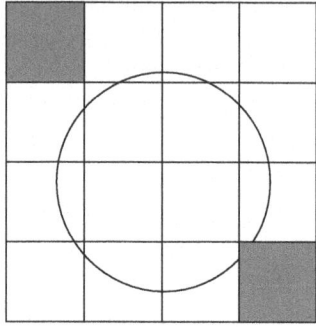

Theorem 5.4.6. *There is a polynomial-time 28-approximation for* MINW-DS *in any block B.*

Proof. Suppose for each node v, $disk_1(v)$ intersects at most α cells e. Then

$$c(\cup_{e \in B} A(e)) \leq \sum_{e \in B} c(A(e)) \leq \sum_{e \in B} 2c(\text{Opt}(e)) \leq 2\alpha c(\text{Opt}(b))$$

where $\text{Opt}(B)$ is the optimal solution for MINW-DS in the block B.

Note that $disk_1(v)$ can intersect at most four horizontal strips and at most four vertical strips, hence at most 16 cells. Furthermore, consider two cells at two ends of a diagonal. Only one of them has its interiors intersecting a disk with radius one (Fig. 5.6). Thus, $\alpha \leq 14$. This means that $\cup_{e \in B} A(e)$ is a 28-approximation for MINW-DS in the block B. □

5.5 6-Approximation

Why and O_1 and O_2 be computed within polynomial-time in the **2-Approximation** for MINW-DS in a cell e? This section first answers this question. To do so, it suffices to study the following problem.

MINW-SENSOR-COVER with Targets in a Strip: Consider a set P of targets lying inside a horizontal strip and a set $\mathcal{D}$ of disks with radius one and centers lying either above or below the strip (Fig. 5.7). Assume every target in P is covered by at least one disk in D. Given disks with a nonnegative weight $c : \mathcal{D} \to R^+$, find the minimum total weight subset of disks covering all targets.

Let

$$\mathcal{D}^+ = \{D \in \mathcal{D} \mid \text{the center of } D \text{ lies above the strip}\}$$

and

$$\mathcal{D}^- = \{D \in \mathcal{D} \mid \text{the center of } D \text{ lies below the strip}\}.$$

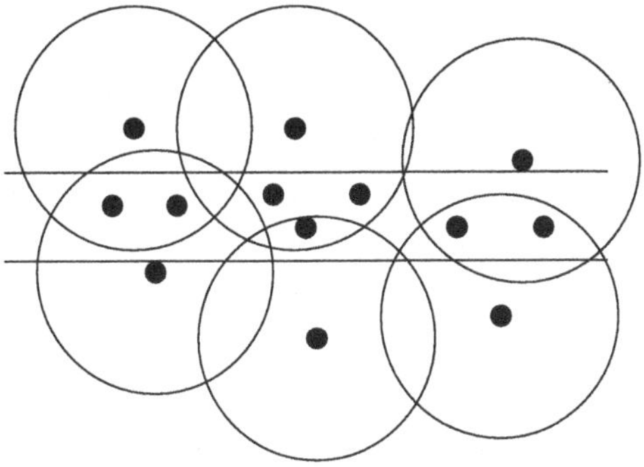

5.5 6-Approximation

Fig. 5.7 Sensor Cover with targets in a strip

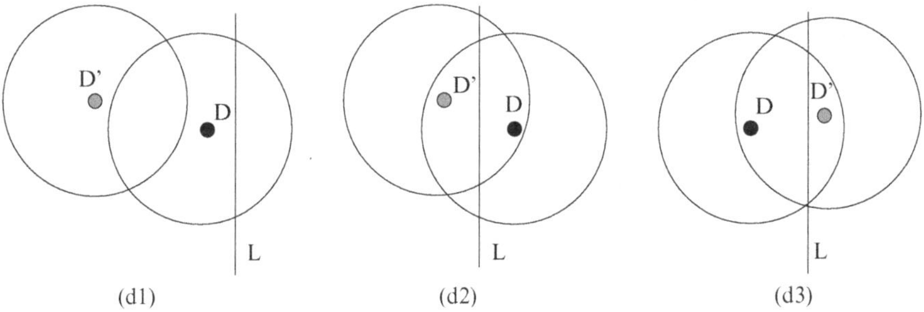

Fig. 5.8 D' is controlled by D at line L

Consider a disk $D \in \mathcal{D}^+$ intersecting a vertical line L. A disk $D' \in \mathcal{D}^+$ is said to be *controlled* by D at L, denoted by $D' \prec_L^+ D$, if one of the following holds (Fig. 5.8):

(d1) D' does not intersect L.
(d2) The lower endpoint of $D' \cap L$ is higher than the lower endpoint of $D \cap L$.
(d3) The lower endpoint of $D' \cap L$ is identical to the lower endpoint of $D \cap L$. But, the center of D' is on the right of the center of D.

Similarly, let $D \in \mathcal{D}^-$ intersect a line L. Then a disk $D' \in \mathcal{D}^-$ is said to be *controlled* by D, denoted by $D' \prec_L^- $, if one of the following holds:

(d1) D' does not intersect L.
(d2) The upper endpoint of $D' \cap L$ is lower than the upper endpoint of $D \cap L$.
(d3) The upper endpoint of $D' \cap L$ is identical to the upper endpoint of $D \cap L$. But, the center of D' is on the right of the center of D.

The following are important properties of controlledness.

Lemma 5.5.1. *Let* $D, D', D'' \in \mathcal{D}^+$ *and* L *a vertical line. If* $D'' \prec_L^+ D'$ *and* $D' \prec_L^+ D$, *then* $D'' \prec_L^+ D$. *Similarly, for* $D, D', D'' \in \mathcal{D}^-$, *if* $D'' \prec_L^- D'$ *and* $D' \prec_L^- D$, *then* $D'' \prec_L^- D$.

Proof. It follows immediately from the definition of controlledness. □

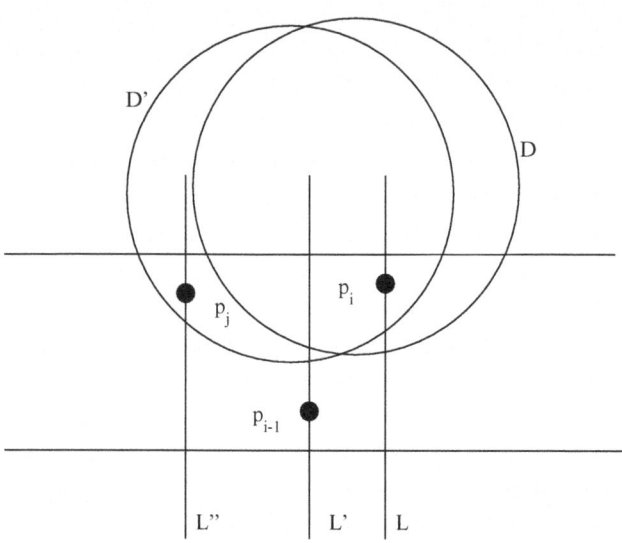

Fig. 5.9 The proof of Lemma 5.5.1

Lemma 5.5.2. *Let D and D' be two disks and L a vertical line. Suppose $D \cap L \neq \emptyset$. Then*

(e1) If $D, D' \in \mathcal{D}^+$, then either $D' \prec_L^+ D$ or $D \prec_L^+ D'$;
(e2) If $D, D' \in \mathcal{D}^-$, then either $D' \prec_L^- D$ or $D \prec_L^- D'$.

Proof. It follows immediately from the definition of controlledness. □

Lemma 5.5.3. *Let L and L' be two vertical lines such that L lies on the right of L'. Then the following holds:*

(L1) Let $D, D \in \mathcal{D}^+$. If $D' \prec_L^+ D$ and $D \prec_{L'}^+ D'$, then $D \cap Q(L') \subseteq D' \cap Q(L')$ where $Q(L')$ is the closed lower-left quarter-plane bounded by the upper boundary of the strip and L'.
(L2) Let $D, D \in \mathcal{D}^-$. If $D' \prec_L^- D$ and $D \prec_{L'}^- D'$, then $D \cap Q'(L') \subseteq D' \cap Q'(L')$ where $Q'(L')$ is the closed upper-left quarter-plane bounded by the lower boundary of the strip and L'.

Proof. For contradiction, suppose (L1) is not true, that is, there exists a point $p \in (D \cap Q(L')) \setminus (D' \cap Q(L'))$. Let L'' be the vertical line passing through p. Then $D' \prec_{L''}^+ D$. Then L'' is on the left of L'. Let A be the lower endpoint of $D \cap L$ and A'' the lower endpoint of $D \cap L''$. Then $D \cap L' \neq \emptyset$. Let A' be the lower endpoint of $D \cap L'$ and B' the lower endpoint of $D' \cap L'$. Then A' is above B'. They both should lie below the line AA'' (Fig. 5.9). Therefore, $D' \cap AA''$ is located inside the segment $[A, A'']$. Let E and F be two endpoints of $D' \cap AA''$. Clearly, $\angle EAF < \angle A''A'A$ and $|EF| \leq |AA''|$. Therefore,

5.5 6-Approximation

$$\text{radius}(D') = \frac{|EF|}{2\sin \angle EB'F} < \frac{|AA''|}{2\sin \angle A''A'A} = \text{radius}(D).$$

a contradiction. □

Now, it is ready to show the following.

Theorem 5.5.4 (Ambühl et al. [2]). MINW-SENSOR-COVER *with targets in a strip can be solved in* $O(m^4 n)$ *where* $n = |P|$ *and* $m = |\mathcal{D}|$.

Proof. First, assume every disk in $\mathcal{D}$ has a positive weight since disks with zero weight can be removed together with targets covered by them at the beginning.

Let $p_1, ..., p_n$ be all points in P in the ordering from left to right. Call as an *upper disk* (*lower disk*) for any disk with center above (below) the strip. A dynamic programming will be employed to find the optimal solution. For simplicity of description, assume the two boundaries of the strip are also two disks with infinite radius and weight zero. They do not cover any point in P. The upper bound is an upper disk, and the lower bound is a lower disk. Note that these two disks do not belong to $\mathcal{D}$. But, the relations $\prec_L^+$ and $\prec_L^+$ can be extended to all upper disks and all lower disks, respectively.

For an upper disks D and a lower disk D' with $D \cup D'$ covering p_i, define by $T_i(D, D')$ the one with the minimum total weight among disk subsets $\mathcal{D}'$ satisfying the following conditions:

1. $\mathcal{D}'$ covers $p_1, \ldots, p_i$.
2. $D, D' \in \mathcal{D}'$.
3. Let L_i be the vertical line passing through p_i. Then D controls every upper disk in $\mathcal{D}'$ at L_i, and D' controls every lower disk in $\mathcal{D}'$ at L_i.

Since two boundaries of the strip have zero weight and cover nothing, for simplicity of the discussion, one assume that they cannot appear in $T_i(D,D') - \{D,D'\}$. In other word, they can play only the role of D or D'.

Let $c(T_i(D,D'))$ be the total weight of disks in $T_i(D,D')$. One claims that the following recursion holds.

$$c(T_i(D,D')) = \min_{D_1, D_2} \{c(T_{i-1}(D_1, D_2)) + c(\{D,D'\} \setminus \{D_1, D_2\})\}, \quad (5.1)$$

where upper disk D_1 and lower disk D_2 are over all possible pairs satisfying conditions:

(c1) $D_1 \cup D_2$ covers p_{i-1}.
(c2) Let L_i be the vertical line passing through p_i. Then $D_1 \prec_{L_i}^+ D$ and $D_2 \prec_{L_i}^- D'$.

To show this claim, at the first choose D_1 to be the upper disk in $T_i(D,D')$ which controls every upper disk in $T_i(D,D')$ at L_{i-1}. By Lemma 5.5.1, such a choice must exist. Similarly, one can choose D_2 to the lower disk in $T_i(D,D')$ which controls

every lower disk in $T_i(D,D')$. By Lemma 5.5.3, $D \cap Q(L_{i-1}) \subseteq D_1 \cap Q(L_{i-1})$ and $D' \cap Q' \subseteq D_2 \cap Q'$. Therefore, $(T_i(D,D') - \{D,D'\}) \cup \{D_1,D_2\}$ covers $p_1,\ldots,p_{i-1}$. Hence,
$$c(T_i(D,D')) - c(\{D,D'\} \setminus \{D_1,D_2\}) \geq c(T_{i-1}(D_1,D_2)),$$
that is,
$$c(T_i(D,D')) \geq \min_{D_1,D_2}(c(T_{i-1}(D_1,D_2)) + c(\{D,D'\} \setminus \{D_1,D_2\})).$$

On the other hand, for any pair $\{D_1,D_2\}$ satisfying (c1) and (c2), $T_{i-1}(D_1,D_2) \cup \{D,D'\}$ covers $p_1,\ldots,p_i$. Moreover, for any upper disk $\hat{D}$ in $T_{i-1}(D_1,D_2)$, one must have $\hat{D} \prec^+_{L_i} D$. Indeed, for contradiction, suppose $D \prec^+_{L_i} \hat{D}$. Then $\hat{D} \neq D_1$ and hence $\hat{D} \in \mathcal{D}$ has a positive weight. By Lemma 5.5.1, $D_1 \prec^+_{L_i} \hat{D}$. Note that by the definition of $T_{i-1}(D_1,D_2)$, $\hat{D} \prec^+_{L_{i-1}} D_1$. By Lemma 5.5.3, then $\hat{D} \cap Q(L_{i-1}) \subseteq D_1 \cap Q(L_{i-1})$, which means that $\hat{D}$ can be deleted from $T_{i-1}(D_1,D_2)$. This contradicts the minimality of $T_{i-1}(D_1,D_2)$.

Similarly, for any lower disk $\hat{D}$ in $T_{i-1}(D_1,D_2)$, one must have $\hat{D} \prec^-_{L_i} D'$. Therefore,
$$c(T_i(D,D')) \leq T_{i-1}(D_1,D_2) + c(\{D,D'\} \setminus \{D_1,D_2\})$$
for any pair $\{D_1,D_2\}$ satisfying (c1) and (c2). Therefore,
$$c(T_i(D,D')) \leq \min_{D_1,D_2}(T_{i-1}(D_1,D_2) + c(\{D,D'\} \setminus \{D_1,D_2\}))$$
for $\{D_1,D_2\}$ over all pairs satisfying (c1) and (c2). Hence, (5.1) holds.

This recursion suggests a dynamic program for computing all $T_i(D,D')$. There are $O(nm^2)$ $T_i(D,D')$'s and each needs to be computed recursively in time $O(m^2)$. Therefore, this dynamic program runs in time $O(nm^4)$. Finally, the minimum weight of subset of disks covering all targets can be computed from $\min_{D,D'} c(T_n(D,D'))$, which requires $O(m^2)$ time. □

By Theorem 5.5.4, O_1 and O_2 in **2-Approximation** for MINW-DS in a cell e can be computed in polynomial time. Hence, a polynomial-time 28-approximation has been obtained for MINW-DS in a block B.

However, an idea motivated from Theorem 5.5.4 may give a big improvement. That is to combine $V_1(e)$ along a horizontal strip and combine $V_2(e)$ along a vertical strip. With such an idea, the approximation performance ratio can be reduced from 28 to 6.

6-Approximation for MINW-DS in a block B.

input a unit disk graph $G = (V,E)$ and a block B.

Let C be the set of cells e in B with $V(e) = V \cap e \neq \emptyset$. Let $H_1,\ldots,H_m$ be horizontal strips and $Y_1,\ldots,Y_m$ vertical strips of B.

5.5 6-Approximation

Let $C' \subseteq C$. For each cell $e \in C'$, choose a vertex $v_e \in V(e)$ and let $U = \{v_e \mid e \in C'\}$. For every subset C' and every U, compute a vertex subset $A(C',U)$ in the following way:

Step 1. Let $Z = V \cap (\cup_{v \in U} \mathrm{disk}_1(v))$. For every $e \in C - C'$, update $V(e) \leftarrow V(e) \setminus Z$.
Step 2. For every cell $e \in C - C'$ and for every choice of $\{p,p',q,q'\} \subseteq V(e)$, let
$V_1(e) = V(e) \cap (\Delta_{\mathrm{south}}(p,p') \cup \Delta_{\mathrm{north}}(q,q'))$ and $V_2(e) = V(e) - V_1(e)$.

   Step 2.1. For each horizontal strip H_i, compute a minimum weight subset $\mathrm{Opt}(H_i)$ of disks with centers lying outside H_i to dominate $(\cup_{e \in H_i \cap (C-C')} V_1(e))$.
   Step 2.2. For each vertical strip Y, compute a minimum weight subset $\mathrm{Opt}(Y_i)$ of disks with centers lying outside Y_i to dominate $(\cup_{e \in Y_i \cap (C-C')} V_2(e))$.
   Step 2.3 Compute $O = (\cup_{i=1}^m \mathrm{Opt}(H_i)) \cup (\cup_{i=1}^m \mathrm{Opt}(Y_i))$.
   Step 2.4 Compute O^* to minimize the total weight $c(O)$ over all possible combinations of $\{p.p',q,q'\}$ for all $e \in C - C'$.

Step 3. Set $A(C',U) = O^* \cup U$.

Finally, compute an $A^* = A(C',U)$ to minimize the total weight $c(A(C',U))$ for C' over all subsets of C and U over all choices of v_e for all $e \in C'$.
output A^*.

Theorem 5.5.5 (Huang et al. [66]). *There exists a 6-approximation for* MINW-DS *in a block B, with running time $n^{O(m^2)}$ where n is the number of nodes v such that $\mathrm{disk}_1(v) \cap B \neq \emptyset$.*

Proof. First, let us estimate the time for computing A^*. There are $O(2^{m^2})$ possible subsets of C, $n^{O(m^2)}$ possible choices of U and $O(n^{4m^2})$ possible combinations of $\{p,p',q,q'\}$ for all cells in $C - C'$. For each combination, computing all $\mathrm{Opt}(H_i)$ and all $\mathrm{Opt}(Y_i)$ needs time $O(n^5)$. Therefore, total computation time is $n^{O(m^2)}$.

Next, estimate the performance ratio. Let Opt be an optimal solution for MINW-DS in the block B. Set $C' = \{e \in C \mid e \cap \mathrm{Opt} \neq \emptyset\}$. For each $e \in C'$, choose a node $v_e \in \mathrm{Opt} \cap e$. Set $U = \{v_e \mid e \in C'\}$ and $Z = V \cap (\cup_{v \in U} \mathrm{disk}_1(v))$. Update $V(e)$ for all $e \in C - C'$ by $V(e) \leftarrow V(e) \setminus Z$. For each $e \in C - C'$, by Lemma 5.4.4, there exists a set $\{p,p',q,q'\}$ of at most four nodes in $V(e)$ such that $V_1(e) = V(e) \cap (\Delta_{\mathrm{south}}(p,p') \cup \Delta_{\mathrm{north}}(q,q'))$ is dominated by $\mathrm{Opt}_1(e) = \mathrm{Opt}(e) \cap (S \cup N)$ and $V_2(e) = V(e) - V_1(e)$ is dominated by $\mathrm{Opt}_2(e) = \mathrm{Opt}(e) \cap (E \cup W)$ where

$$\mathrm{Opt}(e) = \{v \in V - U \mid e \cap \mathrm{disk}_1(v) \neq \emptyset\}.$$

Note that $\cup_{e \in H_i \cap (C - C_i)} \mathrm{Opt}_1(e)$ is a feasible solution for the minimization problem solved at Step 2.1. Therefore,

$$c(\mathrm{Opt}(H_i)) \leq c(\cup_{e \in H_i \cap (C - C_i)} \mathrm{Opt}_1(e)).$$

Fig. 5.10 Each $\text{disk}_1(v)$ intersects at most six strips not containing v

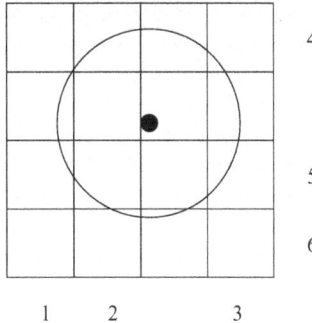

Similarly,
$$c(\text{Opt}(Y_i)) \leq c(\cup_{e \in Y_i \cap (C-C_i)} \text{Opt}_2(e)).$$

Therefore
$$c(O) \leq \sum_{i=1}^{m} c(\cup_{e \in H_i \cap (C-C_i)} \text{Opt}_2(e))$$
$$+ \sum_{i=1}^{m} c(\cup_{e \in Y_i \cap (C-C_i)} \text{Opt}_2(e))$$
$$\leq 6 \cdot c(\text{Opt} - U)$$

since each disk $\text{disk}_i(v)$ can intersect at most six strips which do not contain v (Fig. 5.10). Hence,
$$c(A^*) \leq c(A(C', U)) = c(O^*) + c(U)$$
$$\leq c(O) + c(U)$$
$$\leq 6 \cdot c(\text{Opt} - U) + c(U)$$
$$\leq 6 \cdot c(\text{Opt}).$$
□

5.6 4-Approximation

Zou et al. [134] studied a generalization of MinW-Sensor-Cover with targets in a strip.

> MINW-CHROMATIC-DISK-COVER: Consider m parallel horizontal strips $H_1, \ldots, H_m$ as shown in Fig. 5.11. To have all strips disjoint, assume that each strip has open boundary on the top and close boundary on the bottom. Given a set $\mathcal{R}$ of red disks with radius one, a set $\mathcal{B}$ of blue disk with radius one, a positive weight function $c : \mathcal{R} \cup \mathcal{B} \to R^+$, and a set P of targets points lying in those strips, find the minimum-weight subset of red disks and blue disks such that every target in strip H_i is covered by a chromatic disk, that is, by either a red disk with center lying above H_i or a blue disk lying below H_i.

5.6 4-Approximation

Fig. 5.11 Chromatic disk cover

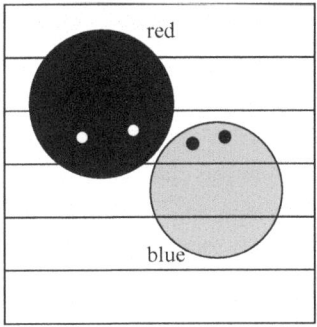

When $m = 1$, MINW-CHROMATIC-DISK-COVER is exactly MINW-SENSOR-COVER with targets in a strip.

Theorem 5.6.1 (Zou et al. [134]). *MINW-CHROMATIC-DISK-COVER can be solved in time $O(nd^{4m})$ where $n = |P|$ and $d = |\mathcal{R} \cup \mathcal{B}|$.*

Proof. Denote $\mathcal{R}_i = \mathcal{R} \cap H_{i-1}$ for $i = 2,\ldots,m$ and $\mathcal{R}_1 = \{\text{disk}_1(v) \in \mathcal{R} \mid v \text{ lies above } H_1\}$. Denote $\mathcal{B}_i = \mathcal{B} \cap H_{i+1}$ for $i = 1,\ldots,m-1$ and $\mathcal{B}_m = \{\text{disk}_1(v) \in \mathcal{B} \mid v \text{ lies below } H_m\}$.

Let $\mathcal{R}_i^+$ be obtained from $\mathcal{R}$ by putting a dumming disk with radius infinite, which is the lower half plane bounded by the upper boundary of H_i. Let $\mathcal{B}_i^+$ be obtained from $\mathcal{B}_i$ by putting a dumming disk with radius infinite, which is the upper half plane bounded by the lower bound of H_i. Consider a $2m$-dimensional vectors $\mathbf{D}$ in

$$\mathcal{S} = \mathcal{R}_1^+ \times \cdots \times \mathcal{R}_m^+ \times \mathcal{B}_1^+ \times \cdots \times \mathcal{B}_m^+.$$

For simplicity, $\mathbf{D}$ is also used to denote the set of components of $\mathbf{D}$. Denote by D_i the ith component of $\mathbf{D}$. Then one has $D_i \in \mathcal{R}_i^+$ and $D_{m+i} \in \mathcal{B}_i^+$ for $1 \leq i \leq m$. Consider $\mathbf{D}, \mathbf{D}' \in \mathcal{S}$. For any vertical line L, $\mathbf{D}'$ is said to be controled by $\mathbf{D}$, written as $\mathbf{D}' \prec_L \mathbf{D}$, if for $1 \leq i \leq m$, $D_i' \prec_L^+ D_i$ and $D_{m+i}' \prec_L^- D - m + i$.

Let $L_1,..,L_k$ be all vertical lines passing through target points in P in the ordering from left to right. Let P_i be the subset of targets in P lying on L_i or on the left of L_i. For any $\mathbf{D} \in \mathcal{S}$, if $\mathbf{D}$ does not cover $P_j - P_{j-1}$, then define $T_j(\mathbf{D}) = $ nil if there does not exist a disk subset $\mathcal{D}'$ satisfying the following conditions:

- $\mathcal{D}'$ is a chromatic disk cover for P_j.
- $\mathbf{D} \subseteq \mathcal{D}'$.
- For any disk $D' \in \mathcal{D}' \cap \mathcal{R}_i$, $D' \prec_{L_j}^+ D_i$.
- For any disk $D' \in \mathcal{D}' \cap \mathcal{B}_i$, $D' \prec_{L_j}^- D_{m+i}$.

If such a disk subset $\mathcal{D}'$ exists, then define $T_j(\mathcal{D})$ to be the one with the minimum total weight.

Since all dumming disks have zero weight and cover nothing, for simplicity of the discussion, one assumes that they cannot appear in $T_j(\mathbf{D}) \setminus \mathbf{D}$.

Now, one claims that for **D** covering $P_j - P_{j-1}$, the following recursion holds.

$$c(T_j(\mathbf{D})) = \min_{\mathbf{D}' \prec_{L_j} \mathbf{D}} \{c(T_{j-1}(\mathbf{D}')) + c(\mathbf{D} \setminus \mathbf{D}')\}. \tag{5.2}$$

To show this claim, one first chooses D'_i to be the disk in $T_j(\mathcal{D}) \cap \mathcal{R}^+_j$ which controls every disk in $T_j(\mathbf{D}) \cap \mathcal{R}^+_j$ at L_{j-1}. By Lemma 5.5.1, such a choice must exists. Similarly, one can choose D'_{m+i} to the disk in $T_j(\mathbf{D})$ which controls every disk in $T_j(\mathbf{D}) \cap \mathcal{B}^+_i$ at line L_{j-1}. Define $\mathbf{D}' = (D'_i, 1 \leq i \leq 2m)$.

By Lemma 5.5.3, $D_i \cap Q_i(L_{j-1}) \subseteq D'_i \cap Q_i(L_{j-1})$ for $1 \leq i \leq m$ where $Q_i(L_j)$ is the close lower-left quarter-plane bounded by L_{j-1} and the upper boundary of H_i. Similarly, $D_{m+i} \cap Q'_i(L_{j-1}) \subseteq D'_{m+i} \cap Q'_i(L_{j-1})$ for $1 \leq i \leq m$. This means that if $T_j(\mathbf{D}) \neq$ nil, then $(T_j(\mathbf{D})) - (\mathbf{D} \setminus \mathbf{D}')$ is a chromatic disk cover for P_{j-1}. Hence,

$$c(T_j(\mathbf{D})) - c(\mathbf{D} \setminus \mathbf{D}') \geq c(T_{j-1}(\mathbf{D}')),$$

that is,

$$c(T_j(\mathbf{D})) \geq \min_{\mathbf{D}' \prec_{L_j} \mathbf{D}} (c(T_{j-1}(\mathbf{D}')) + c(\mathbf{D} \setminus \mathbf{D}')). \tag{5.3}$$

If $T_j(\mathbf{D}) =$ nil, then $c(T_j(\mathbf{B})) = \infty$ and hence (5.3) holds trivially.

On the other hand, for any $\mathbf{D}' \prec \mathbf{D}$, if $T_{j-1}(\mathbf{D}') =$ nil, then $c(T_{j-1}(\mathbf{D}')) = \infty > c(T_j(\mathbf{D}))$. Next, assume that $T_{j-1}(\mathbf{D}') \neq$ nil. Then $T_{j-1}(\mathbf{D}') \cup \mathbf{D}$ is a chromatic cover of P_j.

Moreover, for any disk $\hat{D}$ in $T_{j-1}(\mathbf{D}')$, one must have $\hat{D} \prec^+_{L_j} D_i$ if $\hat{D} \in \mathcal{R}^+_i$ and $\hat{D} \prec^-_{L_j} D_{m+i}$ if $\hat{D} \in \mathcal{B}^+_i$. In fact, for contradiction, suppose $\hat{D} \in \mathcal{R}^+_i$ and $\hat{D}$ is not controlled by D_i. Thus, $\hat{D} \notin \mathbf{D}'$ and hence $c(\hat{D}) > 0$. Moreover, by Lemma 5.5.2, $D_i \prec^+_{L_j} \hat{D}$. By Lemma 5.5.1, $D'_i \prec^+_{L_j} \hat{D}$. By Lemma 5.5.3, $\hat{D} \cap Q_i(L_{j-1}) \subseteq D'_i \cap Q_i(L_{j-1})$. This means that $\hat{D}$ can be deleted from $T_{j-1}(\mathcal{D}')$, contradicting the minimality of $T_{j-1}(\mathbf{D}')$. Similarly, it is also impossible that $\hat{D} \in \mathcal{B}^+_i$ and $\hat{D}$ is not controlled by D_{m+i}.

From above argument, one can see that $T_{j-1}(\mathbf{D}') \cup \mathbf{D}$ satisfies all conditions for above $\mathcal{D}'$. Therefore,

$$c(T_j(\mathbf{D})) \leq c(T_{j-1}(\mathbf{D}' \cup \mathbf{D}))$$
$$= c(T_{j-1}(\mathbf{D}')) + c(\mathbf{D} \setminus \mathbf{D}')$$

for all $\mathbf{D}' \prec_{L_j} \mathbf{D}$. This completes the proof of (5.2).

The recursion (5.2) suggests a dynamic program for computing all $T_j(\mathbf{D})$. There are $O(nd^{2m})$ $T_j(\mathbf{D})$'s, and each needs to be computed recursively in time $O(d^{2m})$. Therefore, this dynamic program runs in time $O(nd^{4m})$. Finally, the minimum weight of subset of disks covering all targets can be computed from $\min_{\mathbf{D} \in S} c(T_k(\mathbf{D}))$, which requires $O(d^{2m})$ time. □

5.6 4-Approximation

4-Approximation for MINW-DS in a block B.
 input a unit disk graph $G = (V, E)$ and a block B.
 Let C be the set of cells e in B with $V(e) = V \cap e \neq \emptyset$. Let $H_1, \ldots, H_m$ be horizontal strips and $Y_1, \ldots, Y_m$ vertical strips of B.
 Let $C' \subseteq C$. For each cell $e \in C'$, choose a vertex $v_e \in V(e)$ and let $U = \{v_e \mid e \in C'\}$. For every subset C' and every U, compute a vertex subset $A(C', U)$ in the following way:

Step 1. Let $Z = V \cap (\cup_{v \in U} \text{disk}_1(v))$. For every $e \in C - C'$, update $V(e) \leftarrow V(e) \setminus Z$.
Step 2. For every cell $e \in C - C'$ and for every choice of $\{p, p', q, q'\} \subseteq V(e)$, let $V_1(e) = V(e) \cap (\Delta_{\text{south}}(p, p') \cup \Delta_{\text{north}}(q, q'))$ and $V_2(e) = V(e) - V_1(e)$.

  Step 2.1. Compute an optimal solution $\text{Opt}(H)$ for MINW-CHROMATIC-DISK-COVER with horizontal strips $H_1, \ldots, H_m$, target set $P = (\cup_{e \in (C-C')} V_1(e))$, red disk set $\mathcal{R} = \{(\text{disk}_1(v), \text{red}) \mid v \in V - U\}$ and blue disk set $\mathcal{B} = \{(\text{disk}_1(v), \text{blue}) \mid v \in V - U\}$.
  Step 2.2. Compute an optimal solution $\text{Opt}(Y)$ for MINW-CHROMATIC-DISK-COVER with vertical strips $Y_1, \ldots, Y_m$, target set $P = (\cup_{e \in (C-C')} V_2(e))$, red disk set $\mathcal{R} = \{(\text{disk}_1(v), \text{red}) \mid v \in V - U\}$ and blue disk set $\mathcal{B} = \{(\text{disk}_1(v), \text{blue}) \mid v \in V - U\}$. (Note: Each target is required to be covered by either a red disk from the left or a blue disk from the right.)
  Step 2.3 Compute $O = \text{Opt}(H) \cup \text{Opt}(Y)$.
  Step 2.4 Compute O^* to minimize the total weight $c(O)$ over all possible combinations of $\{p.p', q, q'\}$ for all $e \in C - C'$.

Step 3. Set $A(C', U) = O^* \cup U$.

Finally, compute an $A^* = A(C', U)$ to minimize the total weight $c(A(C', U))$ for C' over all subsets of C and U over all choices of v_e for all $e \in C'$.
 output A^*.

Theorem 5.6.2 (Zou et al. [134]). *There exists a 4-approximation for* MINW-DS *in a block B, with running time* $n^{O(m^2)}$ *where n is the number of nodes v such that* $\text{disk}_1(v) \cap B \neq \emptyset$.

Proof. First, let us estimate the time for computing A^*. There are $O(2^{m^2})$ possible subsets of C, $n^{O(m^2)}$ possible choices of U and $O(n^{4m^2})$ possible combinations of $\{p, p', q, q'\}$ for all cells in $C - C'$. For each combination, computing all $\text{Opt}(H)$ and all $\text{Opt}(Y)$ needs time $O(n^{4m+1})$. Therefore, total computation time is $n^{O(m^2)}$.

Next, estimate the performance ratio. Let Opt be an optimal solution for MINW-DS in the block B. Set $C' = \{e \in C \mid e \cap \text{Opt} \neq \emptyset\}$. For each $e \in C'$, choose a node $v_e \in \text{Opt} \cap e$. Set $U = \{v_e \mid e \in C'\}$ and $Z = V \cap (\cup_{v \in U} \text{disk}_1(v))$. Update $V(e)$ for all $e \in C - C'$ by $V(e) \leftarrow V(e) \setminus Z$. For each $e \in C - C'$, by Lemma 5.4.4, there exists a set $\{p, p', q, q'\}$ of at most four nodes in $V(e)$ such that $V_1(e) = V(e) \cap (\Delta_{\text{south}}(p, p') \cup$

$\Delta_{\text{north}}(q,q'))$ is dominated by $\text{Opt}_1(e) = \text{Opt}(e) \cap (S \cup N)$ and $V_2(e) = V(e) - V_1(e)$ is dominated by $\text{Opt}_2(e) = \text{Opt}(e) \cap (E \cup W)$ where

$$\text{Opt}(e) = \{v \in V - U \mid e \cap \text{disk}_1(v) \neq \emptyset\}.$$

Let

$$\mathcal{D}_1(e) = \{(\text{disk}_1(v), \text{red}) \mid v \in \text{Opt}_1(v) \cap N\} \cup \{(\text{disk}_1, \text{blue}) \mid v \in \text{Opt}_1(v) \cap S\}.$$

Then $\cup_{e \in (C-C_i)} \mathcal{D}_1(e)$ is a feasible solution for the minimization problem solved at Step 2.1. Therefore,
$$c(\text{Opt}(H)) \leq c(\cup_{e \in (C-C_i)} \mathcal{D}_1(e)).$$

Similarly, let

$$\mathcal{D}_2(e) = \{(\text{disk}_1(v), \text{red}) \mid v \in \text{Opt}_2(v) \cap E\} \cup \{(\text{disk}_1, \text{blue}) \mid v \in \text{Opt}_2(v) \cap W\}.$$

Then $\cup_{e \in (C-C_i)} \mathcal{D}_2(e)$ is a feasible solution for the minimization problem solved at Step 2.2. Therefore,
$$c(\text{Opt}(Y)) \leq c(\cup_{e \in (C-C_i)} \mathcal{D}_2(e)).$$

Therefore

$$\begin{aligned} c(O) &\leq c(\cup_{e \in (C-C_i)} \mathcal{D}_1(e)) \\ &\quad + c(\cup_{e \in (C-C_i)} \mathcal{D}_2(e)) \\ &\leq 4 \cdot c(\text{Opt} - U) \end{aligned}$$

since each disk $\text{disk}_i(v)$ can involve feasible solutions for at most two minimization problems and at each feasible solution $\text{disk}_1(v)$ has at most two copies, one in red and one in blue. Hence,

$$\begin{aligned} c(A^*) &\leq c(A(C', U)) = c(O^*) + c(U) \\ &\leq c(O) + c(U) \\ &\leq 4 \cdot c(\text{Opt} - U) + c(U) \\ &\leq 4 \cdot c(\text{Opt}). \end{aligned}$$

$\square$

5.7 3.63-Approximation

Erlebach and Mihalak [46] studied another generalization of MINW-SENSOR-COVER with targets in a strip.

5.7 3.63-Approximation

Fig. 5.12 Multi-strips

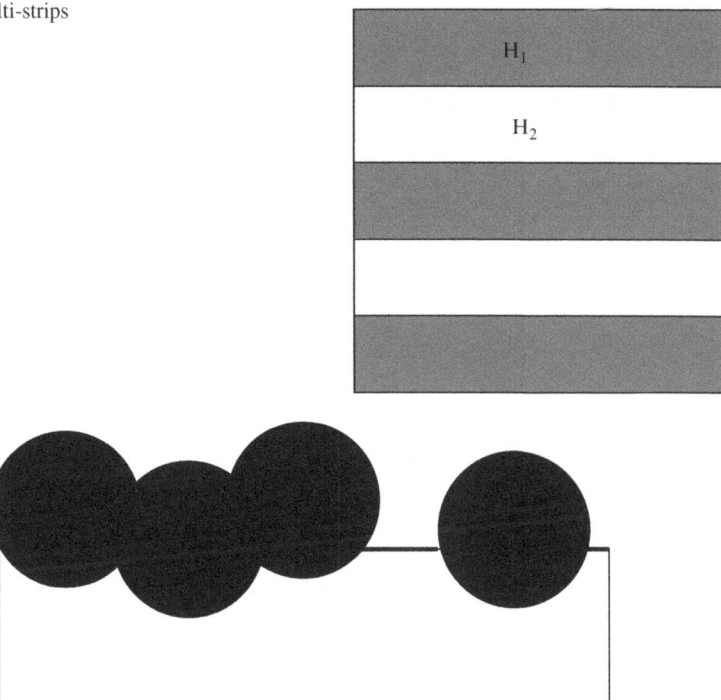

Fig. 5.13 The upper envelope

MINW-SENSOR-COVER with Targets in Multi-Strips: Consider m parallel horizontal strips $H_1, \ldots, H_m$ in a block as shown in Fig. 5.12. To have all strips disjoint, assume that each strip has open boundary on the top and close boundary on the bottom. Given a set $\mathcal{D}$ of n disks with radius one and a positive weight function $c : \mathcal{D} \to R^+$, and a set P of k targets points lying in strips $H_1 \cup H_3 \cup \cdot \cup H_{2\lceil m/2 \rceil - 1}$, find the minimum weight subset of disks such that every target in strip H_i is covered by a disk with center lying outside of H_i.

Erlebach and Mihalak [46] transformed this problem to a shortest path problem. To explain this transformation, the first is to study an optimal solution Opt for MINW-SENSOR-COVER with targets in multi-strips.

Consider a strip H_i for some odd i, $1 \leq i \leq m$. A disk in Opt is called an *upper disk* with respect to H_i if its center lies above H_i. For simplicity of discussion, the half plane above H_i is also considered as an upper (dumming) disk with respect to H_i. All upper disks with respect to H_i form an *upper area* of H_i. The boundary of this area is called the *upper envelope* of H_i (Fig. 5.13). Similarly, one may define lower disks, the lower dumming disks the lower area and the lower envelope of H_i.

Lemma 5.7.1. *An upper disk* $\text{disk}_1(v)$ *with respect to* H_i *can appear in the upper envelope of* H_i *at most once. If an upper disk* $\text{disk}_1(u)$ *is on the left of another upper disk* $\text{disk}_1(v)$ *on the upper envelope of* H_i*, then the center* u *is on the left of the center* v*.*

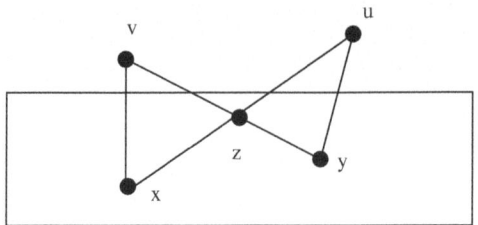

Fig. 5.14 The proof of Lemma 5.7.1

Similarly, a lower disk $disk_1(v)$ *with respect to* H_i *can appear in the lower envelope of* H_i *at most once. If a lower disk* $disk_1(u)$ *is on the left of another lower disk* $disk_1(v)$ *on the lower envelope of* H_i, *then the center u is on the left of the center v.*

Proof. For contradiction, suppose the upper disk $disk_1(u)$ is on the left of the upper disk $disk_1(v)$ on the upper envelope of strip H_i, but the center u is on the right of the center v. Choose a point x from $circle_1(u)$ appearing on the upper envelope and a point y from $circle_1(v)$ appearing on the upper envelope. Then segments ux and vy intersect, say at z. Since x and y appear in the upper envelope, one must have $|ux| < |vx|$ and $|vy| < |uy|$. Therefore, $|ux| + |vy| < |vx| + |uy|$. However, $|vx| < |vz| + |zx|$ and $|uy| < |uz| + |zy|$. Hence $|vx| + |uy| < |vz| + |zx| + |uz| + |zy| = |ux| + |vy|$, a contradiction. Therefore, the second sentence is true (Fig. 5.14).

The first sentence is a corollary of the second sentence. In fact, if $disk_1(v)$ appears twice on the upper envelope, then between two appearances, there must exist another disk $disk_1(u)$ appearing. This means that v is on the left of u and also on the right of u, a contradiction. □

A *corner* of the upper envelope of H_i is an intersection point of two upper disks on the envelope. A *corner* of the lower envelope of H_i is an intersection point of two lower disks on the envelope. A *sweep line* L_i for H_i is a vertical line that starts from a position on the left of all disks in $\mathcal{D}$ and moves to right until a position on the right of all disks in $\mathcal{D}$. L_i's movement is discrete. Each intermediate position of L_i must pass through a corner on either the upper envelope or the lower envelope. Each of such positions is denoted by a quadruple $(d_1,d_2;d_3,d_4)$ with either $d_1 = d_2$ or $d_3 = d_4$ where d_1,d_2 are upper disks and d_3,d_4 are lower disks. If $d_1 \neq d_2$, then L_i passes through the intersection point of d_1 and d_2 on the upper envelope. If $d_3 \neq d_4$, then L_i passes through the intersection point of d_3 and d_4 on the lower envelope. For the initial and the end position of L_i, $d_1 = d_2$ is the upper dumming disk and $d_3 = d_4$ is the lower dumming disk.

A *move* of L_i is from its current position $(d_1,d_2;d_3,d_4)$ to an adjacent position on the right. If $d_1 = d_2$, this adjacent position is either $(d_2,d;d_4,d_4)$ or $(d_1,d_1;d_4,d)$. In this case, one says that the disk d_3 leaves L_i and the disk d enters. If $d_3 = d_4$, then the right adjacent position is either $(d_2,d;d_4,d_4)$ or $(d_2,d_2;d_4,d)$. In this case, one says that the disk d_1 leaves and the disk d enters (Fig. 5.15).

5.7 3.63-Approximation

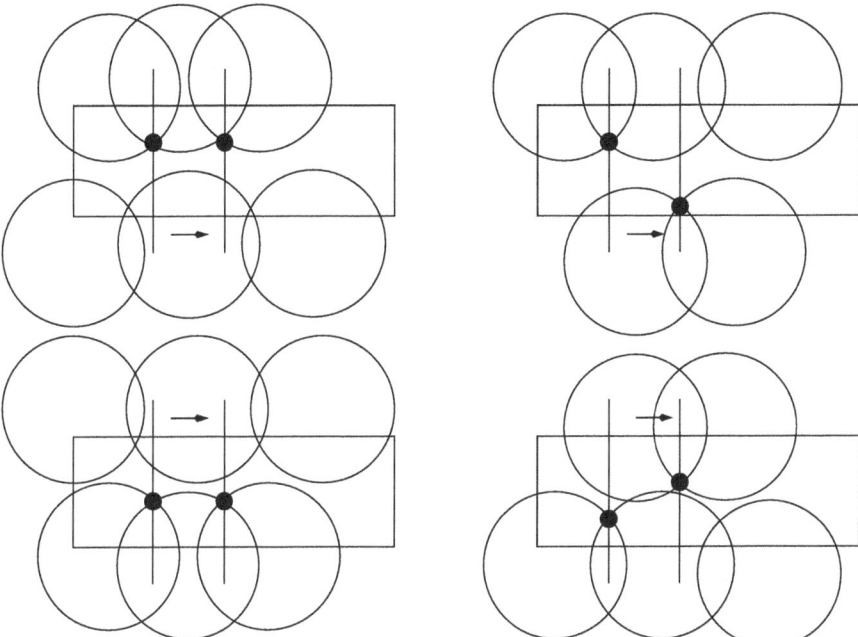

Fig. 5.15 Four possibilities for a sweep line to move to right adjacent position

Note that every disk in Opt must appear in the upper or lower envelope for some strip H_i and the weight of Opt equals the total weight of disks appearing on envelopes. The sweep line is used to calculate this total weight so that one can turn the sweeping process into a shortest path of a graph. There is one trouble if the sweep line is doing counting individually. Each disk may appear in the lower envelope of strip H_i and also in the upper envelope of H_{i+2}. If every sweep line does individual counting, then the total weight is not exactly $c(\mathrm{Opt})$. Therefore, all sweep line may be required to do a combined action and to employ a technique for avoiding the double counting.

A *configuration* of sweep lines consists of positions of sweep lines $L_1, L_3, \ldots, L_{2\lceil m/2 \rceil - 1}$ at a moment. A *legal move* from a configuration A to another configuration B consists of exactly one move of one sweep line which is required to satisfy the following constraint:

(m1) In this move, if a lower disk d leaves line L_i, then the disk d should be already passed by L_{i+2}. Otherwise, L_i has to wait for L_{i+2} to pass the disk d.
(m2) In this move, if an upper disk d leaves line L_i, then the disk d should be already passed by L_{i-2}. Otherwise, L_i has to wait for L_{i-2} to pass the disk d.

Here, by a disk passed by a sweep line at position $(d_1, d_2; d_3, d_4)$, one means that if d is an upper disk, then the center of d is not on the right of the center of d_2; if d is a lower disk, then the center of d is not on the right of the center of d_4.

Why can this constraint eliminate the double counting? It is because one can set counting rule as follows: The counting is performed only on a newly entered disk in each move. This means that when a disk d enters a configuration in a move, the weight of d is counted if the configuration does not contain d, and the weight of d is not counted if the configuration already contains d.

Can this constraint stop the moving of configuration? The answer is no because if the moving stops, then every sweep line is waiting for another line to pass a disk. Then there are two cases.

Case 1. There are two sweep lines L_i and L_{i+2} such that L_i waits for L_{i+2} to pass a disk d and L_{i+2} waits for L_i to pass another disk d'. In this case, L_i should be in position $(d_1, d_2; d, d_4)$ and L_{i+2} should be in position $(d', d'_2; d'_3, d'_4)$. Since L_i waits for L_{i+2} to pass d, the center of d is on the right of the center of d'_2 and hence by Lemma 5.7.1, the center of d is on the right of the center of d'. Similarly, the center of d' should be on the right of the center of d, a contradiction.

Case 2. Case 1 does not occur. If L_i waits for L_{i+2}, then L_{i+2} waits for L_{i+4}, etc. However, since the number of strips if finite, this process cannot go forever, a contradiction. If L_i waits for L_{i-2}, then L_{i-2} waits for L_{i-4}. This process cannot go forever, neither.

Next, an auxiliary graph $G(\text{Opt})$ can be constructed to turn the moving of configuration into a shortest path. All possible configurations form all vertices. There exists an arc (A, B) from a configuration A to another configuration B if and only if B can be reached from A through a legal move. The start vertex s is the configuration consisting of all sweep lines on the left of all disks in $\mathcal{D}$. The target vertex is the configuration consisting of all sweep lines on the right of all disks in $\mathcal{D}$.

Now, it is ready to show the following.

Theorem 5.7.2 (Eriksson et al. [45]). MINW-SENSOR-COVER *with targets in multi-strips can be solved in time* $O(n^{3(m+1)})$.

Proof. Use $\mathcal{D}$ instead of Opt to construct the sweep line positions, configurations, legal move of configurations, and graph $G(\mathcal{D})$ by following the same rules in the construction of graph $G(\text{Opt})$, except an additional requirement for a sweep line move: during the move, targets between two positions should be covered by two envelope. Then $G(\mathcal{D})$ contains $G(\text{Opt})$ as a subgraph, and the shortest path from configuration s to configuration t would give an optimal solution for SENSOR COVER with targets in multi-strips. Since each sweep line position is determined by three disks, each sweep line has at most $O(n^3)$ positions. Hence, the number of configurations is at most $O(n^{3m/2})$ for even m and $O(n^{3(m+1)/2})$ for odd m. Therefore, computing the shortest path in $G(\mathcal{D})$ takes time $O(n^{3m})$ for even m and $O(n^{3(m+1)})$ for odd m. □

Based on Theorem 5.7.2, Willson et al. [114] constructed a polynomial-time 3.63-approximation for MINW-DS in a block B. Their main idea is motivated from the following observation. To construct an approximation solution, MINW-DS in a block B is divided into four problems, two on horizontal strips and two on vertical strips. Consider the two on horizontal strips. One is on strips $H_1, H_3, \ldots$.

5.7 3.63-Approximation

Fig. 5.16 $disk_1(v)$ involves only one problem if v is nearly at the central of H_i

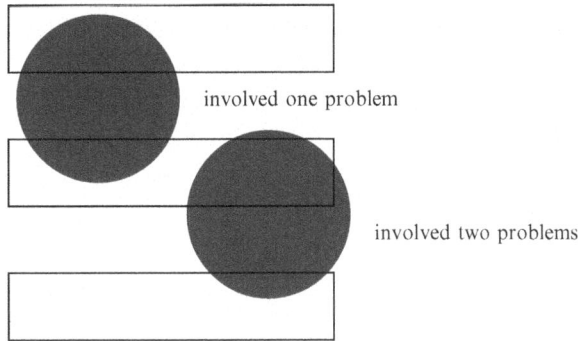

involved one problem

involved two problems

The other one is on strips $H_2, H_4, \ldots$ Suppose $disk_1(v)$ with $v \in H_i$. Then $disk_1(v)$ will involve only one problem which is on strips H_{i-1} and H_{i+1} if v lies nearly at the central of H_i. This means that in average, $disk_1(v)$ involves less than two problems. How to take advantage of this average estimation? Shifting the partition on the block B is a traditional technique (Fig. 5.16).

3.63-Approximation for MINW-DS in a block.
input a unit disk graph $G = (V, E)$ and a block B.
Put the block B at the position with $(0,0)$ as its left-lower corner. Let $P(a,b)$ be a grid with cell size $\mu \times \mu$ and the left-lower corner
$(-a m \mu / q, -b m \mu / q)$ such that the block B is covered where $\mu = \frac{\sqrt{2}}{2}$.
$A \leftarrow nil$ ($c(nil) = \infty$);
for $a = 0$ **to** $q - 1$ **do**
    for $b = 0$ **to** $q - 1$ **do begin**
        compute $A(a,b)$ with procedure $A(a,b)$;
        if $c(A) > c(A(a,b))$
            then $A \leftarrow A(a,b)$;
    end-for;
output A.

Procedure $A(a,b)$
    input a unit disk graph $G = (V, E)$ and a block B. Use grid $P(a,b)$ partition the block B into cells.
Let C be the set of cells e in B with $V(e) = V \cap e \neq \emptyset$. Let $H_1, \ldots, H_m$ be horizontal strips and $Y_1, \ldots, Y_m$ vertical strips of B. Define

$$H_{odd} = H_1 \cup H_3 \cup \cdots \cup H_{2\lceil m/2 \rceil - 1},$$
$$H_{even} = H_2 \cup H_4 \cup \cdots \cup H_{2\lfloor m/2 \rfloor},$$
$$Y_{odd} = Y_1 \cup Y_3 \cup \cdots \cup Y_{2\lceil m/2 \rceil - 1},$$
$$Y_{even} = Y_2 \cup Y_4 \cup \cdots \cup Y_{2\lfloor m/2 \rfloor}.$$

Let $C' \subseteq C$. For each cell $e \in C'$, choose a vertex $v_e \in V(e)$ and let $U = \{v_e \mid e \in C'\}$. For every subset C' and every U, compute a vertex subset $A(C',U)$ in the following way:

Step 1. Let $Z = V \cap (\cup_{v \in U} \text{disk}_1(v))$. For every $e \in C - C'$, update $V(e) \leftarrow V(e) \setminus Z$.

Step 2. For every cell $e \in C - C'$ and for every choice of $\{p,p',q,q'\} \subseteq V(e)$, let $V_1(e) = V(e) \cap (\Delta_{\text{south}}(p,p') \cup \Delta_{\text{north}}(q,q'))$ and $V_2(e) = V(e) - V_1(e)$.

> Step 2.1. Compute an optimal solution $\text{Opt}(H_{\text{odd}})$ for SENSOR-COVER with targets in multi-strips and with horizontal strips H_{odd}, target set $P = (\cup_{e \in (C-C') \cap H_{\text{odd}}} V_1(e))$, disk set $\mathcal{D} = \{\text{disk}_1(v) \mid v \in V - U\}$.
>
> Step 2.2. Compute an optimal solution $\text{Opt}(H_{\text{even}})$ for SENSOR-COVER with targets in multi-strips and with horizontal strips H_{even}, target set $P = (\cup_{e \in (C-C') \cap H_{\text{even}}} V_1(e))$, disk set $\mathcal{D} = \{\text{disk}_1(v) \mid v \in V - U\}$.
>
> Step 2.3. Compute an optimal solution $\text{Opt}(Y_{\text{odd}})$ for SENSOR-COVER with targets in multi-strips and with vertical strips Y_{odd}, target set $P = (\cup_{e \in (C-C') \cap Y_{\text{odd}}} V_2(e))$, disk set $\mathcal{D} = \{\text{disk}_1(v) \mid v \in V - U\}$.
>
> Step 2.4. Compute an optimal solution $\text{Opt}(Y_{\text{even}})$ for SENSOR-COVER with targets in multi-strips with vertical strips Y_{even}, target set $P = (\cup_{e \in (C-C') \cap Y_{\text{even}}} V_2(e))$, disk set $\mathcal{D} = \{\text{disk}_1(v) \mid v \in V - U\}$.
>
> Step 2.5 Compute $O = \text{Opt}(H_{\text{odd}}) \cup \text{Opt}(H_{\text{even}}) \cup \text{Opt}(Y_{\text{odd}}) \cup \text{Opt}(Y_{\text{even}})$
>
> Step 2.6 Compute O^* to minimize the total weight $c(O)$ over all possible combinations of $\{p.p',q,q'\}$ for all $e \in C - C'$.

Step 3. Set $A(C',U) = O^* \cup U$.

Finally, compute an $A(a,b) = A(C',U)$ to minimize the total weight $c(A(C',U)$ for C' over all subsets of C and U over all choices of v_e for all $e \in C'$.

output $A(a,b)$.

Theorem 5.7.3 (Willson et al. [114]). *There exists a 3.63-approximation for MINW-DS in a block B, with running time $n^{O(m^2)}$ where n is the number of nodes v such that $\text{disk}_1(v) \cap B \neq \emptyset$.*

Proof. First, let us estimate the time for computing $A(a,b)$. There are $O(2^{m^2})$ possible subsets of C, $n^{O(m^2)}$ possible choices of U and $O(n^{4m^2})$ possible combinations of $\{p,p',q,q'\}$ for all cells in $C - C'$. For each combination, computing $\text{Opt}(H_{\text{odd}})$, $\text{Opt}(H_{\text{even}})$, $\text{Opt}(Y_{\text{odd}})$ and $\text{Opt}(Y_{\text{even}})$ needs time $O(n^{4m+1})$. Therefore, total computation time is $n^{O(m^2)}$.

Next, estimate the performance ratio. Let Opt be an optimal solution for MINW-DS in the block B. Set $C' = \{e \in C \mid e \cap \text{Opt} \neq \emptyset\}$. For each $e \in C'$, choose a node $v_e \in \text{Opt} \cap e$. Set $U = \{v_e \mid e \in C'\}$ and $Z = V \cap (\cup_{v \in U} \text{disk}_1(v))$. Update $V(e)$ for all $e \in C - C'$ by $V(e) \leftarrow V(e) \setminus Z$. For each $e \in C - C'$, by Lemma 5.4.4, there exists a set $\{p,p',q,q'\}$ of at most four nodes in $V(e)$ such that $V_1(e) = V(e) \cap (\Delta_{\text{south}}(p,p') \cup$

5.7 3.63-Approximation

$\Delta_{\text{north}}(q,q'))$ is dominated by $\text{Opt}_1(e) = \text{Opt}(e) \cap (S \cup N)$ and $V_2(e) = V(e) - V_1(e)$ is dominated by $\text{Opt}_2(e) = \text{Opt}(e) \cap (E \cup W)$ where

$$\text{Opt}(e) = \{v \in V - U \mid e \cap \text{disk}_1(v) \neq \emptyset\}.$$

Then that $\cup_{e \in (C-C_i) \cap H_{\text{odd}}} \text{Opt}_1(e)$ is a feasible solution for the minimization problem solved at Step 2.1. Therefore,

$$c(\text{Opt}(H_{\text{odd}})) \leq c(\cup_{e \in (C-C_i) \cap H_{\text{odd}}} \text{Opt}_1(e)). \tag{5.4}$$

Similarly,

$$c(\text{Opt}(H_{\text{even}})) \leq c(\cup_{e \in (C-C_i) \cap H_{\text{even}}} \text{Opt}_1(e)). \tag{5.5}$$

$$c(\text{Opt}(Y_{\text{odd}})) \leq c(\cup_{e \in (C-C_i) \cap Y_{\text{odd}}} \text{Opt}_2(e)). \tag{5.6}$$

$$c(\text{Opt}(H_{\text{odd}})) \leq c(\cup_{e \in (C-C_i) \cap Y_{\text{even}}} \text{Opt}_2(e)). \tag{5.7}$$

For every $v \in \text{Opt}$, define

$$\tau_1(v;a,b) = \begin{cases} 1 & \text{if } \text{disk}_1(v) \text{ intersects three horizontal strips,} \\ 2 & \text{otherwise,} \end{cases}$$

and

$$\tau_2(v;a,b) = \begin{cases} 1 & \text{if } \text{disk}_1(v) \text{ intersects three vertical strips,} \\ 2 & \text{otherwise.} \end{cases}$$

Then $\text{disk}_1(v)$ involves $\tau_1(v;a,b)$ of two equations (5.4) and (5.5), and $\tau_2(v;a,b)$ of two equations (5.6) and (5.7), Therefore

$$c(O) \leq c(\cup_{e \in (C-C_i) \cap H_{\text{odd}}} \text{Opt}_1(e)) + c(\cup_{e \in (C-C_i) \cap H_{\text{even}}} \text{Opt}_1(e))$$
$$+ c(\cup_{e \in (C-C_i) \cap Y_{\text{odd}}} \text{Opt}_2(e)) + c(\cup_{e \in (C-C_i) \cap Y_{\text{even}}} \text{Opt}_2(e))$$
$$\leq \sum_{v \in \text{Opt}-U} c(v)(\tau_1(v;a,b) + \tau_2(v:a,b)).$$

Hence,

$$c(A(a,b)) \leq c(A(C',U)) = c(O^*) + c(U)$$
$$\leq c(O) + c(U)$$
$$\leq \sum_{v \in \text{Opt}} c(v)(\tau_1(v;a,b) + \tau_2(v:a,b)).$$

For any $v \in \text{Opt}$, note that for any fixed b, there exist at least $\lfloor \frac{3\mu-2}{\mu/q} \rfloor$ values of a such that $\tau_1(v;a,b) = 1$, and for any fixed a, there exist at least $\lfloor \frac{3\mu-2}{\mu/q} \rfloor$ values of b such that $\tau_2(v;a,b) = 1$. Therefore,

$$A \leq \frac{1}{q^2} \sum_{a=0}^{q-1} \sum_{b=0}^{q-1} \sum_{v \in \text{Opt}} c(v)(\tau_1(v;a,b) + \tau_2(v;a,b))$$

$$\leq \left(4 - 2\frac{\lfloor (3-2\sqrt{2})q \rfloor}{q}\right) \cdot c(\text{Opt}).$$

As $q \to \infty$, $\frac{\lfloor (3-2\sqrt{2})q \rfloor}{q}$ goes to $3 - 2\sqrt{2}$. Since $4 - 2(3 - 2\sqrt{2}) < 3.63$, there exists a fixed q such that $(4 - 2\frac{\lfloor (3-2\sqrt{2})q \rfloor}{q}) < 3.63$. □

Chapter 6
Coverage

The only difference between suicide and martyrdom is press coverage.

CHUCK PALAHNIUK

6.1 Motivation and Overview

A classic type of resource management problem is as follows: Given a certain amount of resource and a set of users, find an assignment of resource to maximize the number of satisfied users. The maximum lifetime coverage is such a classic type of problem in wireless sensor networks.

When a very large number of sensors are randomly deployed into a certain region possibly by an aircraft to monitor a certain set of targets, usually, there are a lot of redundant sensors. A better usage of those redundant sensors is to schedule active/sleep time of sensors to increase the lifetime of the system.

A simple scheduling is to divide sensors into disjoint subsets, each of which fully covers all targets, called a *sensor cover* [18, 80].

SENSOR-COVER-PARTITION: Given n targets $r_1, \ldots, r_n$ and m sensors $s_1, \ldots, s_m$, each covering a subset of targets, find the maximum number of disjoint sensor covers.

This problem is NP-hard. Various heuristics and approximation algorithms have been given in [11, 13, 96]. In general, there is no polynomial-time $(-\varepsilon)\ln n$-approximation for any $\varepsilon > 0$ unless $NP \subseteq DTIME(n^{O(\log \log n)})$ [48] and there exists polynomial-time $O(\log n)$-approximation [6, 80]. But, there is an open problem in a special case.

Open Problem 6.1.1. *Suppose all sensors are uniform, that is, they have the same sensing radius. It is unknown whether a polynomial-time constant-approximation exists or not.*

When the sensor set and the target set are identical, SENSOR-COVER-PARTITION becomes the following domatic partition problem.

MAX#DS: Given a graph $G = (V, E)$, partition the vertex set V into maximum number of disjoint dominating sets.

In general graph, there is no polynomial-time $(1 - \varepsilon)\ln n$-approximation unless $NP \subseteq DTIME(n^{O(\log \log n)})$ and there exists polynomial-time $O(\log n)$-approximation for MAX#DS [48]. However, for unit disk graphs, there is a polynomial-time constant-approximation [86].

For this type of scheduling, the sensor is activated only once, that is, once the sensor is activated, it keeps active until it dies.

Cardei et al. [15] found that it is possible to increase the lifetime if each sensor is allowed to alternate between active and sleeping states. An example can be found in Chap. 1. The model is also better supported by an interesting fact discovered in [64] that putting a sensor alternatively in active and sleeping states in a proper way may double its lifetime since the battery could be recovered in a certain level during sleeping. The formulation of this model is as follows.

MAX-LIFETIME COVERAGE: Given n targets $t_1, \ldots, t_n$ and m sensors $s_1, \ldots, s_m$, each covering a subset of targets, find a family of sensor cover $S_1, \ldots, S_p$ with time lengths $t_1, \ldots, t_p$ in $[0, 1]$, respectively, to maximize $t_1 + \cdots + t_p$ subject to that the total active time of every sensor is at most 1.

This is still an NP-hard problem. Cardei [15] formulated it as a 0-1 integer programming and designed a heuristic without guaranteed theoretical bound. Berman et al. [6, 7] first designed an approximation algorithm for MAX-LIFETIME COVERAGE with theoretical bound. They showed that there exists a polynomial-time approximation for MAX-LIFETIME COVERAGE with performance ratio $O(\log n)$ where n is the number of sensors. By employing Garg–Könemann theorem [55], Berman et al. reduced MAX-LIFETIME COVERAGE to the following:

MINW-SENSOR-COVER: Consider n targets $t_1, \ldots, t_n$ and m sensors $s_1, \ldots, s_m$, each covering a subset of targets. Given a weight function on sensors $c : \{s_1, \ldots, s_m\} \to R^+$, find the minimum total weight sensor cover.

They showed that if MINW-SENSOR-COVER has a polynomial-time ρ-approximation, then MAX-LIFETIME COVERAGE has a polynomial-time $(1 + \varepsilon)\rho$-approximation for any $\varepsilon > 0$. Note that MINW-SENSOR-COVER is equivalent to MINW-SENSOR-COVER. Therefore, it has a polynomial-time $(1 + \log n)$-approximation. Hence, MAX-LIFETIME SENSOR COVER has a polynomial-time $O(\log n)$-approximation. Actually, the first one who found the application of Garg-Könemann theorem in study of lifetime maximization type of problems is Calnescu et al. [12].

Ding et al. [34] noted that all results in Chap. 5 about MINW-DS can be extended to MINW-SENSOR-COVER in the case that all sensors and targets lie in the Euclidean plane and all sensors have the same covering radius. Therefore, they proved that in this case, MAX-LIFETIME COVERAGE has polynomial-time 3.63-approximation.

Du et al. [37] extended this approach to study the coverage problem with connectivity requirement. They constructed a polynomial-time constant-approximation in geometric case and $O(\log n)$-approximation in general case. However, many maximum lifetime coverage with connectivity requirement are still open. The following is an example.

Open Problem 6.1.2. *Does* MAX#CDS *have a polynomial-time constant-approximation in unit disk graphs?*

6.2 Max-Lifetime Connected Coverage

As described in the previous section, the method of Garg and Könemann [55] plays an important role in design of constant-approximation for various problems on the maximum lifetime coverage. In this section, we introduce it through the work of Du et al. [37].

Du et al. [37] studied a quite general model of wireless sensor networks which was previously studied by Zhang and Li [126]. In this model, each sensor has two modes, active mode and sleep mode, and the active mode has two phases, the full-active phase and the semi-active phase. A full-active sensor can sense, transmit, receive, and relay the data packets. A semi-active sensor cannot sense data packets, but it can transmit, receive, and relay data packets. Usually, a sensor in the full-active phase consumes more energy than in the semi-active phase.

Sensors are often randomly deployed into hostile environment, such as battlefield and inaccessible area with chemical or nuclear pollution, so that recharging batteries of sensors is a mission impossible. Assume the battery of each sensor contains a certain amount of energy, say unit amount. Then the lifetime of each sensor depends on energy consumption.

Du et al. [37] studied the following problem:

MAX-LIFETIME CONNECTED-COVERAGE with two active phases: Given a set of targets and a set of sensors with two active phases, find an active/sleeping schedule for sensors to maximize the system lifetime where the network system is said to be *alive* if the following conditions are satisfied:

(A1) Every target is monitored by a full-active sensor.
(A2) All (full-/semi-) active sensors induce a connected subgraph.

They studied this problem with the primal-dual method of Garg and Könemann [55].

Let S be the set of all sensors. Assume all sensors are uniform, that is, they have the same communication radius R_c, the same sensing radius R_s, the same full-active energy consumption u of unit time and the same semi-active energy consumption v of unit time. Also, assume $u \geq v$. A pair p of sets is called an *active sensor set pair* if $p = (p_1, p_2)$ where p_1 is a set of full-active sensors and p_2 is a set of semi-active sensors with $p_1 \cap p_2 = \emptyset$. For any active sensor set pair p, define

$$a_{s,p} = \begin{cases} u & \text{if } s \in p_1, \\ v & \text{if } s \in p_2, \\ 0 & \text{otherwise.} \end{cases}$$

Suppose $\mathcal{C}$ is the collection of all active sensor set pairs satisfying conditions (A1) and (A2). Then MAX-LIFETIME CONNECTED COVERAGE with two active phases can be formulated as the following linear programming:

$$\max \sum_{p \in \mathcal{C}} x_p$$

$$\text{subject to} \quad \sum_{p \in \mathcal{C}} a_{s,p} x_p \leq 1 \quad \text{for } s \in S$$

$$x_p \geq 0 \quad \text{for } p \in \mathcal{C}.$$

Its dual is as follows.

$$\min \sum_{s \in S} y_s$$

$$\text{subject to} \quad \sum_{s \in S} a_{s,p} y_s \geq 1 \quad \text{for } p \in \mathcal{C},$$

$$y_s \geq 0 \quad \text{for } s \in S.$$

Motivated from the work of Garg and Könemann [55], Du et al. [37] designed the following primal-dual algorithm.

Primal-Dual Algorithm DPWW

Initially, choose $x_p = 0$ for all $p \in \mathcal{C}$ and $y_s = \delta$ for all $s \in S$ where δ is a positive constant which will be determined later.

In each iteration, carry out the following steps until $(y_s, s \in S)$ becomes dual feasible, that is, all constrains in dual linear programming are satisfied:

Step 1. Compute a ρ-approximation solution p^* for

MINW-CSC with two active phases:

$$\min_{p \in \mathcal{C}} \sum_{s \in S} a_{s,p} y_s.$$

Step 2. Compute a solution s^* for

$$\max_{s \in S} a_{s,p^*}.$$

6.2 Max-Lifetime Connected Coverage

Step 3. Update x_p and y_s as follows:

(B1) x_p does not change for $p \neq p^*$, and

$$x_{p^*} \leftarrow x_{p^*} + \frac{1}{a_{s^*,p^*}}.$$

(B2) y_s does not change for $s \notin p_1^* \cup p_2^*$, and

$$y_s \leftarrow y_s \left(1 + \theta \frac{a_{s,p^*}}{a_{s^*,p^*}}\right)$$

for $s \in p_1^* \cup p_2^*$ where θ is a constant chosen later.

The following lemmas give two important properties at the end of above algorithm.

Lemma 6.2.1. *At the end of Primal-Dual Algorithm DPWW, $(x_p, p \in C)$ may not be a primal-feasible solution. However, $(x_p/\tau, p \in C)$ is a primal-feasible solution where $\tau = \frac{(v/u)\ln\frac{1+\theta}{v\delta}}{\ln(1+\theta v/u)}$.*

Proof. Note that when y_s gets updated, the following facts must hold:

(a) $(y_s, s \in S)$ is not dual feasible.
(b) $s \in p_1^* \cup p_2^*$.

It follows immediately from (a) that $\sum_{s \in S} a_{s,p^*} y_s < 1$, which together with (b) yields that $y_s < 1/v$ before y_s receives any value change. After y_s is updated, we have

$$y_s < \left(1 + \theta \frac{a_{s,p^*}}{a_{s^*,p^*}}\right)/v \leq (1+\theta)/v.$$

Therefore, at the end of Primal-Dual Algorithm DPWW, $y_s < (1+\theta)/v$.

Now, consider a constraint in the primal linear programming,

$$\sum_{p \in C} a_{s,p} x_p \leq 1,$$

which may not be satisfied after x_p is updated. If updating x_p increases the value of $\sum_{p \in C} a_{s,p} x_p$ by adding $\frac{a_{s,p^*}}{a_{s^*,p^*}}$, then the value of y_s is increased by multiplying a factor $1 + \theta \frac{a_{s,p^*}}{a_{s^*,p^*}}$. Note that the value of $\frac{a_{s,p^*}}{a_{s^*,p^*}}$ has only two possibilities, v/u and 1. Suppose $\frac{a_{s,p^*}}{a_{s^*,p^*}}$ takes value v/u for k times and 1 for ℓ times. Then the value of $\sum_{p \in C} a_{s,p} x_p$ receives an increase in $k(v/u) + \ell$ and

$$(1 + \theta v/u)^k (1+\theta)^\ell \leq \frac{1+\theta}{v\delta}$$

since initially $y_s = \delta$. Moreover, initially, $\sum_{p\in C} a_{s,p} x_p = 0$. Thus, at the end of Primal-Dual Algorithm DPWW, the value of $\sum_{p\in C} a_{s,p} x_p$ is $k(v/u)+\ell$. The maximum value of $k(v/u)+\ell$ can be obtained from the following linear programming with respect to k and ℓ:

$$\max\ k(v/u)+\ell$$
$$\text{subject to}\ k\ln(1+\theta v/u)+\ell\ln(1+\theta) \leq \ln\frac{1+\theta}{v\delta}$$
$$k \geq 0, \ell \geq 0.$$

By theory of the linear programming, the maximum value of objective function can always be achieved by some extreme point. For above one, the feasible domain has three extreme points

$$(0,0),\ \left(0, \frac{\ln\frac{1+\theta}{v\delta}}{\ln(1+\theta)}\right),\ \left(\frac{\ln\frac{1+\theta}{v\delta}}{\ln(1+\theta v/u)}, 0\right).$$

Their objective function values are

$$0,\ \frac{\ln\frac{1+\theta}{v\delta}}{\ln(1+\theta)},\ \frac{v}{u}\cdot\frac{\ln\frac{1+\theta}{v\delta}}{\ln(1+\theta v/u)},$$

respectively. Note that $\frac{z}{\ln(1+\theta z)}$ is strictly monotone decreasing for $z \leq 1$. Thus,

$$0 < \frac{\ln\frac{1+\theta}{v\delta}}{\ln(1+\theta)} < \frac{v}{u}\cdot\frac{\ln\frac{1+\theta}{v\delta}}{\ln(1+\theta v/u)}.$$

Hence, at the end of Primal-Dual Algorithm DPWW,

$$\sum_{p\in C} a_{s,p} x_p \leq \tau = \frac{v}{u}\cdot\frac{\ln\frac{1+\theta}{v\delta}}{\ln(1+\theta v/u)}.$$

Therefore,

$$\sum_{p\in C} a_{s,p} x_p / \tau \leq 1.$$

□

Lemma 6.2.2. *At the end of Primal-Dual Algorithm DPWW,*

$$\sum_{p\in C} x_p/\tau \geq \frac{\ln(v|S|\delta)^{-1}}{\tau\theta\rho}\cdot\text{opt}_{\text{lcc}}$$

6.2 Max-Lifetime Connected Coverage

where opt_{lcc} is the objective function value of optimal solution for MAX-LIFETIME CONNECTED COVERAGE with two active phases and $\tau = (v/u)\log_{1+\theta v/u}\frac{1+\theta}{\delta v}$.

Proof. Denote by $x_p(0)$ the initial value of x_p and by $y_s(0)$ the initial value of y_s. Denote by $x_p(i)$ and $y_s(i)$, respectively, the values of x_p and y_s after the ith iteration. Denote by $s^*(i)$ and $p^*(i)$, respectively, the values of s^* and p^* in the ith iteration. Furthermore, denote $X(i) = \sum_{p \in C} x_p(i)$ and $Y(i) = \sum_{s \in S} y_s(i)$. Then, for $i \geq 1$, one has

$$Y(i) = \sum_{s \in S} y_s(i-1) + \theta \frac{1}{a_{s^*(i),p^*(i)}} \sum_{s \in S} a_{s,p^*(i)} y_s(i-1)$$

$$\leq Y(i-1) + \theta(X(i) - X(i-1)) \rho \min_{p \in C} \sum_{s \in S} a_{s,p} y_s(k-1).$$

Thus,

$$Y(i) \leq Y(0) + \theta \rho \sum_{k=1}^{i} ((X(k) - X(k-1)) \min_{p \in C} \sum_{s \in S} a_{s,p} y_s(k-1).$$

By the duality theory of linear programming, opt_{lcc} is also the objective function value of optimal solution for the dual linear programming. Therefore,

$$\text{opt}_{lcc} = \min_{y_s} \frac{\sum_{s \in S} y_s}{\min_{p \in C} \sum_{s \in S} a_{s,p} y_s},$$

where the minimization is subject to $y_s \geq 0$ for $s \in S$. Hence,

$$\min_{p \in C} \sum_{s \in S} a_{s,p} y_s(k-1) \leq \frac{Y(k-1)}{\text{opt}_{lcc}}.$$

Therefore,

$$Y(i) \leq |S|\delta + \frac{\theta \rho}{\text{opt}} \sum_{k=1}^{i} (X(k) - X(k-1)) Y(k-1).$$

Define

$$w(0) = |S|\delta$$

and

$$w(i) = |S|\delta + \frac{\theta \rho}{\text{opt}} \sum_{k=1}^{i} (X(k) - X(k-1)) w(k-1).$$

It is easy to prove by induction on i that $Y(i) \leq w(i)$. Moreover,

$$w(i) = \left(1 + \frac{\theta \rho}{\text{opt}_{lcc}} (X(i) - X(i-1))\right) w(i-1)$$

$$\leq e^{\frac{\theta \rho}{\text{opt}_{lcc}}(X(i) - X(i-1))} w(i-1)$$

$$\leq e^{\frac{\theta\rho}{\text{opt}_{\text{lcc}}}X(i)}w(0)$$
$$= e^{\frac{\theta\rho}{\text{opt}_{\text{lcc}}}X(i)}|S|\delta.$$

Suppose Primal-Dual Algorithm DPWW stops at the mth iteration. Then $Y(m) \geq 1/v$. Hence

$$1/v \leq Y(m) \leq w(m) \leq |S|\delta e^{\frac{\theta\rho}{\text{opt}_{\text{lcc}}}X(m)}.$$

Therefore,

$$\frac{\text{opt}_{\text{lcc}}}{X(m)/\tau} \leq \frac{\tau\theta\rho}{\ln(v|S|\delta)^{-1}}. \qquad \square$$

Theorem 6.2.3 (Du et al. [37]). *If* MINW-CSC *with two active phases has a polynomial-time ρ-approximation, then* MAX-LIFETIME CONNECTED COVERAGE *with two active phases has a polynomial-time $\rho(1+\varepsilon)$-approximation for any $\varepsilon > 0$.*

Proof. Choose $\delta = (1+\theta)((1+\theta)|S|)^{-\theta}/v$. Note that

$$\frac{\ln\frac{1+\theta}{\delta v}}{\ln(\delta v|S|)^{-1}} = \frac{1}{1-\theta},$$

and $(1+\theta v/u)^{u/(v\theta)+1} > e$ implies $\ln(1+\theta v/u) > \frac{v\theta}{u+v\theta}$. Thus,

$$\frac{\tau\theta\rho}{\ln(v|S|\delta)^{-1}} = \frac{\theta\rho}{(1-\theta)\ln(1+\theta v/u)} \leq \rho \cdot \frac{1+\theta v/u}{1-\theta}.$$

Choose θ such that

$$\frac{1+\theta v/u}{1-\theta} < 1+\varepsilon.$$

Then

$$\frac{\text{opt}}{\sum_{p\in C}x_p/\tau} \leq (1+\varepsilon)\rho.$$

To estimate the running time of Primal-Dual Algorithm DPWW, let p^* be a polynomial time ρ-approximation solution for MINW-CSC with Two Active Phases. Note that every iteration can be carried out in polynomial-time. Therefore, it suffices to estimate the number of iterations. Note that at each iteration, at least one of y_s has its value increased. In the proof of Lemma 6.2.1, it is already proved that at the end of the algorithm, each y_s has its value increased by multiplying at most $\log_{1+\theta v/u}\frac{1+\theta}{\delta v}$. Therefore, the number of iterations is at most

$$|S|\log_{1+\theta v/u}\frac{1+\theta}{\delta v} = \frac{|S|\theta\ln((1+\theta)|S|)}{\ln(1+\theta v/u)} = O(|S|\log|S|),$$

where $\delta v = (1+\theta)((1+\theta)|S|)^{-\theta}$ and θ is fixed as ε is fixed. $\square$

6.3 Domatic Partition

In Chap. 5, it has been shown that there exists a polynomial-time 3.63-approximation for MINW-DS. This result can be extended to the following problem.

MINW-SENSOR-COVER: Consider a set of targets and a set of sensors lying in the Euclidean plane. Suppose all sensors have the same sensing radius R_s, but may have different weights. The problem is to find the minimum weight subset of sensors for covering all targets.

Therefore, the following holds.

Theorem 6.2.4 (Du et al. [37]). MAX-LIFETIME CONNECTED COVERAGE *with Two Active Phases has polynomial-time* $(7.105+\varepsilon)$*-approximations for any* $\varepsilon > 0$ *when all targets and all sensors lie in the Euclidean plane and all sensors are uniform with* $R_c \geq 2R_s$.

Proof. Let Opt$_{\text{CSC}}$ be the optimal solution for MINW-CSC with two active phases. Compute a polynomial-time 3.63-approximation solution A for MINW-SENSOR-COVER with weight $y_s u$ for each sensor s. Then

$$\sum_{s \in A} y_s u \leq 3.63 \cdot \text{opt}_{\text{CSC}},$$

where opt$_{\text{CSC}}$ is the objective function value of Opt$_{\text{CSC}}$. Since $R_c \geq 2R_s$, every sensor in A is adjacent to some sensor in Opt$_{\text{CSC}}$. This means that Opt$_{\text{CSC}} \cup A$ induces a connected subgraph and hence Opt$_{\text{CSC}}$ contains the set of Steiner nodes in a feasible solution for NODE-WEIGHTED STEINER TREE on the terminal set A. Now, find a polynomial-time 3.475-approximation solution B for NODE-WEIGHTED STEINER TREE with weight $y_s v$ for each sensor s. Then

$$\sum_{s \in B} y_s v \leq 3.475 \cdot \sum_{s \in Opt_{\text{CSC}}} y_s v \leq 3.475 \cdot \text{opt}_{\text{CSC}}.$$

Therefore,

$$\sum_{s \in A} y_s u + \sum_{s \in B} y_s v \leq 7.105 \cdot \text{opt}_{\text{CSC}}. \qquad \square$$

6.3 Domatic Partition

So far, the best known constant-approximation for MAX#DS in unit disk graphs is designed also using grid partition, however with a new technique. Let us start to introduce a problem on sensor-cover-partition with a separating line.

SENSOR-COVER-PARTITION with Separating Line: Let L be a horizontal line. Given a set T of targets above L and a set S of sensors with sensing radius one below L, assume that every target is covered by at least one sensor. The problem is to find the maximum number of disjoint sensor covers. (A sensor cover is a subset of sensors covering all targets.)

Let $\delta(S,T) = \min_{t \in T} |\{s \in S \mid t \in \text{disk}_1(s)\}|$ where $\text{disk}_1(s)$ denotes the disk with radius one and the center s. Call as the *skyline* the part, above line L, of envelope of disks $\text{disk}_1(s)$ for all $s \in S$. Let S' be the set of those sensors s such that $\text{circle}_1(s)$ has a piece appearing in the skyline where $\text{circle}_1(s)$ denotes the circle with radius one and the center s. By Lemma 5.7.1 S' lines up from right to left by following their pieces on the skyline. For any $t \in T$, denote $C_{S'}(t) = \text{disk}_1(t) \cap S'$. The following properties are important.

Lemma 6.3.1. *Let s_1, s_2, s_3 be three sensors in S with $s_1.x \leq s_2.x \leq s_3.x$ where $s_i.x$ denotes the x-coordinate of point s_i. Suppose there exists a target t such that $t \in \text{disk}_1(s_1) \cap \text{disk}_1(s_3)$ but $t \notin \text{disk}_2(s_2)$. Then $\text{up}(L) \cap \text{disk}_1(s_2) \subseteq \text{up}(L) \cap (\text{disk}_1(s_1) \cup \text{disk}_1(s_3))$ where $\text{up}(L)$ denotes the half plane above the horizontal line L and $\text{circle}_1(s_2)$ cannot appear in the skyline.*

Proof. It is trivial in the case that $s_1.x = s_2.x$ or $s_2.x = s_3.x$. Thus, we next assume $s_1.x < s_2.x < s_3.x$. For contradiction, suppose there exists a point $p \in \text{up}(L) \cap \text{disk}_1(s_2)$ but $p \notin \text{up}(L) \cap (\text{disk}_1(s_1) \cup \text{disk}_1(s_3))$. Note that $t \in \text{disk}_1(s_1) \cap \text{disk}_1(S_3)$ implies that for any point $q \in \text{up}(L)$ with $q.x = t.x$ and $q.y \leq t.y$, $q \in \text{disk}_1(s_1) \cap \text{disk}_1(S_3)$. Moreover, $t \notin \text{disk}_1(s_2)$ implies that for any $q \in \text{up}(L) \cap \text{disk}_1(s_2)$ with $q.x = t.x$ and $q.y < t.y$ and hence $q \in \text{disk}_1(s_1) \cap \text{disk}_1(S_3)$. It follows that $p.x \neq t.x$. Hence $p.x < t.x$ or $p.x > t.x$. First, consider the case that $p.x < t.x$. In this case, two segments ps_2 and ts_1 must intersect at a point o. Note that $|ps_2| < |ps_1|$ and $|ts_2| > |ts_1|$. Hence, $|ps_2| + |ts_1| < |ps_1| + |ts_2|$. However, by the property of the triangle,
$$|po| + |os_1| \geq |ps_1|$$
and
$$|to| + |os_2| \geq |ts_2|.$$
Therefore
$$|ps_2| + |ts_1| = |po| + |os_2| + |to| + |os_1| \geq |ps_1| + |ts_2|,$$
a contradiction. Similarly, a contradiction can result from the case that $p.x > t.x$.

Note that $\text{circle}_1(s_2) \cap \text{up}(L)$ cannot intersect $\text{up}(L) \cap (\text{circle}_1(s_1) \cup \text{circle}_1(s_3))$. In fact, if they have an intersection point p, then a contradiction can still result from the above argument by noting that the argument still works when $|ps_2| = |ps_1|$. So, $\text{up}(L) \cap \text{disk}_1(s_2)$ is contained strictly inside of $\text{up}(L) \cap (\text{disk}_1(s_1) \cup \text{disk}_1(s_3))$. Hence, $\text{circle}_1(s_2)$ cannot appear in the skyline. □

Lemma 6.3.2. *For any $t \in T$, $C_{S'}(t)$ is a nonempty contiguous subset of the ordered set S'.*

Proof. Suppose $s_1, s_2, s_3 \in S'$ with $s_1.x \leq s_2.x \leq s_3.x$. If $s_1, s_3 \in C_{S'}(t)$ and $s_2.x \notin C_{S'}(t)$, then by Lemma 6.3.1, $s_2 \notin S'$, a contradiction. □

6.3 Domatic Partition

Lemma 6.3.3. *Suppose T' is a subset of targets, satisfying a property that for any two distinct targets $t,t' \in T'$, $C_S(t) \not\subseteq C_S(t')$. Then for any two distinct $t,t' \in T'$, $C_{S'}(t) \cap C_{S'}(t') \neq \emptyset$ implies that $C_{S'}(t)$ contains an endpoint of $C_{S'}(t')$.*

Proof. The lemma holds trivially in the case that $C_{S'}(t)$ is not contained in $C_{S'}(t)$. So, we next assume $C_{S'}(t) \subseteq C_{S'}(t)$. By the assumption on T', there exists $s \in C_S(t) \setminus C_S(t')$. Let s_r and s_l be the right endpoint and the left endpoint of $C_{S'}(t')$. Let $s' \in C_{S'}(t) \cap C_{S'}(t')$. Next, consider two cases.

Case 1. $s_l.x \leq s.x \leq s_r.x$. Note that t' is contained in $\text{disk}_1(s_r)$ and $\text{disk}_1(s_l)$ but not contained in $\text{disk}_1(s)$. By Lemma 6.3.1, $t \in \text{up}(L) \cap \text{disk}_s \subseteq \text{up}(L) \cap (\text{disk}_1(s_l) \cup \text{disk}_1(s_r))$. Therefore, $s_l \in C_{S'}(t)$ or $s_r \in C_{S'}(t)$.

Case 2. $s_l.x > s.x$ or $s_r.x < s.x$. Note that $s_l.x \leq s'.x \leq s_r.x$. For contradiction, suppose t is contained by neither $\text{disk}_1(s_l)$ nor $\text{disk}_1(s_r)$. In the case that $s.x < s_l.x$, t is contained by $\text{disk}_1(s)$ and $\text{disk}_1(s')$, but not contained by $\text{disk}_1(s_l)$. By Lemma 6.3.1, $s_l \notin S'$, a contradiction. Similarly, a contradiction can result from the case that $s_r.x < s.x$. □

Now, it is ready to show the following.

Theorem 6.3.4. *There is a polynomial-time algorithm which can find at least $\delta(S,T)/4$ disjoint sensor covers.*

Proof. Consider the following algorithm.

The DomPart Algorithm.
input: a sensor set S and a target set T.
$j \leftarrow 0$;
$E \leftarrow S$;
while E is a set cover **do begin**
1. $j \leftarrow j+1$;
2. $T' \leftarrow T$;
    while there exist $t,t' \in T'$ such that $C_E(t) \subseteq C_E(t')$
    do $T' \leftarrow T' \setminus \{t'\}$;
3. Let $E' \subseteq E$ contribute the skyline of disks at E;
4. Find a maximal subset T'' of T' such that $C_{E'}(t)$ for $t \in T''$ are disjoint;
5. $A_j = \{\text{two endpoints of } C_{E'}(t) \mid t \in T''\}$;
6. $E \leftarrow E \setminus A_j$;
end-while
output: $A_1, A_2, \ldots, A_j$.

First, we show that each A_i for $i = 1, \ldots, j$ is a sensor cover. In fact, for each $t'' \in T''$, A_i contains two endpoints of $C_{E'}(t'')$ and hence t'' is covered by A_i. For $t' \in T' \setminus T''$, there exists $t'' \in T''$ such that $C_{E'}(t') \cap C_{E'}(t'') \neq \emptyset$. By Lemma 6.3.3, $C_{E'}(t')$ contains an endpoint of $C_{E'}(t'')$ and hence t' is covered by A_i. For $t \in T \setminus T'$, there exists $t' \in T'$ such that $C_E(t') \subseteq C_E(t)$. So, there exists $t'' \in T''$ such that $C_E(t)$ contains an endpoint of $C_{E'}(t'')$ and hence t is covered by A_i.

Next, we show that at the end of the jth iteration, $|C_E(t)| \geq \delta(S,T) - 4j$ for every $t \in T$. To do so, let E_j denote the E at the end of the jth iteration. Suppose this inequality holds at the end of the $(j-1)$th iteration, that is, $|C_{E_{j-1}}(t)| \geq \delta(S,T) - 4(j-1)$ for all $t \in T$. We show that $|C_{E_j}(t)| \geq \delta(S,T) - 4j$ for all $t \in T$.

In the jth iteration, for $t'' \in T''$, two endpoints of $C_{E'}(t'')$ are deleted from E_{j-1} and hence

$$|C_{E_j}(t'')| \geq |C_{E_{j-1}}(t'')| - 2 > \delta(S,T) - 4j.$$

For $t' \in T' \setminus T''$, if $C_{E'}(t')$ contains an endpoint of $C_{E'}(t'')$ for $t'' \in T''$, then by Lemma 6.3.3, $C_{E'}(t'')$ must contain an endpoint of $C_{E'}(t')$. Thus, there are at most two such t'''s because all $C_{E'}(t'')$ for $t'' \in T''$ are disjoint. This means that

$$|C_{E_j}(t')| \geq |C_{E_{j-1}}(t')| - 4 \geq \delta(S,T) - 4j.$$

For $t \in T \setminus T'$, there exists $t' \in T'$ such that $C_{E_{j-1}}(t') \subseteq C_{E_{j-1}}(t)$. This relationship is preserved in the algorithm, that is, $C_{E_j}(t') \subseteq C_{E_j}(t)$. Therefore,

$$|C_{E_j}(t)| \geq |C_{E_j}(t')| \geq \delta(S,T) - 4j.$$

It follows immediately from this inequality that at the end of The DomPart Algorithm, $j \geq \delta(S,T)/4$. □

With Theorem 6.3.4, Pandit et al. [86] constructed an algorithm for MAX#DS in unit disk graphs as follows.

Put input unit disk graph $G = (V,E)$ into a square and partition the square with a grid of cells with diameter one (or say, diagonal length one). A cell is called a *heavy* cell if it contains at least $\delta/14$ nodes where $\delta^{\min}$ is the minimum node degree of G. A cell is *light* if it is not heavy. For each node v in a light cell, $\text{disk}_1(v)$ intersects at most 14 cells, at least one of which contains at least $\delta^{\min}/14$ nodes adjacent to v. Choose such a heavy cell σ and put v to T^σ, say that v belongs to σ. Let $S^\sigma = \sigma \cap V$. Consider S^σ as a sensor set and T^σ as a target set. Then the following lemma gives an important fact.

Lemma 6.3.5. *If for every heavy cell σ, S^σ can be partitioned into k sensor covers for T^σ, then G has k disjoint dominating sets.*

Proof. Choose a sensor cover A^σ for each heavy cell σ. Let A be the union of A^σ for σ over all heavy cells. Then A is a dominating set because each A^σ dominates not only all nodes in T^σ, but also dominates all nodes in S^σ. □

For each heavy cell σ, partition T^σ into four parts $(T^\sigma_{\text{north}}, T^\sigma_{\text{south}}, T^\sigma_{\text{east}}, T^\sigma_{\text{west}})$ where T^σ_{north} consists of nodes lying above the line through the upper bound of σ, T^σ_{south} consists of nodes lying below the line through the lower bound of σ, T^σ_{east} consists of nodes lying in the right of the line through the right bound of σ, and T^σ_{west} consists of nodes lying in the left of the line through the left bound of σ. When two parts are available for a node v in T^σ, v can arbitrarily choose one of them as its home. Corresponding these four parts, partition S^σ also into four parts $(S^\sigma_{\text{north}}, S^\sigma_{\text{south}}, S^\sigma_{\text{east}}, S^\sigma_{\text{west}})$ by independently and randomly distributing each node into these four parts.

6.4 Min-Weight Dominating Set

Now, solve SENSOR-COVER-PARTITION with separation line on four inputs $(S^\sigma_{\text{north}}, T^\sigma_{\text{north}})$, $(S^\sigma_{\text{south}}, T^\sigma_{\text{south}})$, $(S^\sigma_{\text{east}}, T^\sigma_{\text{east}})$, and $(S^\sigma_{\text{west}}, T^\sigma_{\text{west}})$. Combine those solutions into k disjoint dominating sets of G where

$$k = \min\{\delta(S^\sigma_{\text{south}}, T^\sigma_{\text{south}}), \delta(S^\sigma_{\text{east}}, T^\sigma_{\text{east}}), \delta(S^\sigma_{\text{west}}, T^\sigma_{\text{west}}) \mid \text{all heavy cells } \delta\}.$$

Next, we show that $k \geq \delta^{\min}/112$ with a quite high probability.

Note that for each $t \in T^\sigma$, $|\sigma \cap \text{disk}_1(t)| \geq \delta^{\min}/14$ and the probability of at least one of two nodes in $\sigma \cap \text{disk}_1(t)$ distributed in the part containing t is $3/4$. By Chernoff bound, the probability of at least $\delta^{\min}/56$ nodes in $\sigma \cap \text{disk}_1(t)$ distributed in the part containing t is at least $1 - e^{-\delta^{\min}/112}$.

Note that for each heavy cell σ, there are at most 20 cells within distance one to σ. So, there are at most 20 light cells which contain a node belonging to σ. Hence, $|T^\sigma| \leq (20/14)\delta^{\min}$. Thus, the probability of the following held is at least $1 - (20/14)\delta^{\min} e^{-\delta^{\min}/112}$:

$$\min(\delta(S^\sigma_{\text{south}}, T^\sigma_{\text{south}}), \delta(S^\sigma_{\text{east}}, T^\sigma_{\text{east}}), \delta(S^\sigma_{\text{west}}, T^\sigma_{\text{west}})) \geq \delta^{\min}/56.$$

Since the number of heavy cells cannot be bounded by $O(\delta^{\min})$, it is hard to estimate the probability of $k \geq \delta^{\min}/56$. Thus, it requires more efforts on distribution of each element of S^σ in order to establish a solution of the following problem.

Open Problem 6.3.6. *Is there a polynomial-time algorithm which produces $\Omega(\delta^{\min})$ disjoint dominating sets for G with high probability?*

6.4 Min-Weight Dominating Set

Pandit et al. [86] gave an interesting idea to construct approximation algorithms for MINW-DS using algorithm for MAX#DS.

Consider the following LP-relaxation of MINW-DS.

$$\min \sum_{i \in V} c_i x_i$$

$$\text{subject to} \quad \sum_{i \in \text{disk}_1(j)} x_i \geq 1 \text{ for all } j \in V$$

$$x_i \geq 0 \text{ for all } i \in V.$$

Let $(x_i^*, i \in V)$ be an optimal solution of this LP. Denote $n = |V|$. Let

$$\bar{x}_i = \begin{cases} 0 & \text{if } x_i^* \leq 1/2n \\ \frac{k}{2n} & \text{if } \frac{k-1}{2n} < x_i^* \leq \frac{k}{2n}. \end{cases}$$

Lemma 6.4.1. *The following holds:*

(1) For $j \in V$, $\sum_{i \in \mathrm{disk}_1(j)} \bar{x}_i \geq 1/2$.
(2) $\sum_{i \in V} c_i \bar{x}_i \leq 2 \cdot \mathrm{opt}_{\mathrm{WDS}}$ where opt_{eds} is the objective function value of an optimal solution for MinW-DS.

Proof. Since $|V \cap \mathrm{disk}(j)| \leq n$, there are at most n x_i^* are rounded down to 0. Therefore,

$$\sum_{i \in \mathrm{disk}_1(j)} \bar{x}_i \geq 1 - n \cdot \frac{1}{2n} = 1/2.$$

This means that (1) holds. For (2), note that

$$\sum_{i \in V} c_i \bar{x}_i \leq 2 \sum_{i \in \mathrm{disk}_1(j)} c_i x_i^* \leq 2 \cdot \mathrm{opt}_{\mathrm{WDS}}. \qquad \square$$

Construct a set P by making $2n \cdot \bar{x}_j$ copies of node j for each $j \in V$. Suppose each copy of j has the same weight as that of j.

Lemma 6.4.2. $c(P) \leq 4n \cdot \mathrm{opt}_{\mathrm{WDS}}$.

Proof. By Lemma 6.4.1, $c(P) = 2n \cdot \sum_{i \in V} c_i \cdot \bar{x}_i \leq 4n \mathrm{opt}_{\mathrm{WDS}}$. $\qquad \square$

Lemma 6.4.3. $\delta(P, V) \geq n$.

Proof. By Lemma 6.4.1, $\sum_{i \in \mathrm{disk}_1(j)} \bar{x}_i \geq 1/2$. Thus, for each $j \in V$, $|P \cap \mathrm{disk}_1(j)| = 2n \sum_{i \in \mathrm{disk}_1(j)} \bar{x}_i \geq n$. $\qquad \square$

Suppose there is an algorithm which can produce at least $\delta(P,V)/C$ sensor cover packing $A_1, \ldots, A_t$ ($t \geq n/C$) for sensor set P and target set V. Then there exists A_i such that

$$c(A_i) \leq \frac{c(P)}{t} \leq \frac{C \cdot c(P)}{n} \leq \frac{4Cn \cdot \mathrm{opt}_{\mathrm{WDS}}}{n} = 4C \cdot \mathrm{opt}_{\mathrm{WDS}}.$$

This means that the following holds.

Theorem 6.4.4. *If there is a polynomial-time algorithm for SENSOR-COVER-PARTITION which can produce $\delta(P,V)/C$ sensor covers for sensor set P and target set V, then there is a polynomial-time $4C$-approximation for MinW-DS.*

Chapter 7
Routing-Cost Constrained CDS

> *If you do not have brains you follow the same route twice.*
> GREEK PROVERB

7.1 Motivation and Overview

Consider a graph as shown in Fig. 7.1. $C = \{4,5,6\}$ is a minimum CDS. $G = (V,E)$. The minimum routing between nodes 1 and 3 not through C is 1-2-3 and through D is 1-4-5-6-2, which is significantly longer than 1-2-3. This example indicates a problem about CDS that while CDS is introduced to save resources in wireless networks, routing cost and communication delay may be increased.

To keep communication delay within certain limit, Kim et al. [67] constructed a diameter-constrained CDS in the unit disk graph, which has diameter within a constant factor from the diameter of input graph, and meanwhile, the size within a constant factor from the unconstrained minimum CDS.

Motivated from a work of Wu et al. [117], Ding et al. [31] tried to keep the minimum routing cost by studying the problem of constructing the minimum CDS C under the following constraint:

(ROC0) For any pair of nodes u and v, every shortest path between u and v has all intermediate nodes in C.

They showed that MIN-CDS with constraint (ROC0) is polynomial-time solvable. However, the size of (ROC0)-constrained CDS is too big. Therefore, Ding et al. [32] relaxed constraint (ROC0) to the following:

(ROC1) For any pair of nodes u and v, there exists a shortest path with all intermediate nodes in C, connecting u and v.

Willson et al. [115] gave another motivation. Consider Fig. 7.1 again. Through CDS $\{4,5,6\}$, all routing paths from node 1 to node 3, from node 1 to node 9, from node 7 to node 3, and from node 7 to node 9 must pass through road 4-5-6. This

Fig. 7.1 Routing through CDS

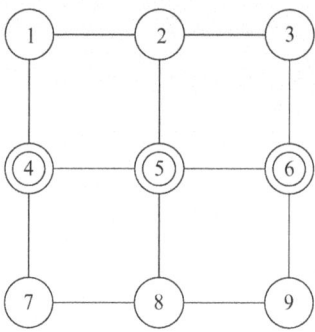

makes road 4-5-6 very crowded. To improve the load balance, they also studied MIN-CDS with constraint (ROC1) and proved NP-hardness of this problem and gave a polynomial-time $O(n)$-approximation with distributed construction.

Ding et al. [32] showed that computing the minimum (ROC1)-constrained CDS in general graph has no polynomial-time $\rho \ln \Delta$-approximation for $0 < \rho < 1$ unless $NP \subseteq DTIME(n^{O(\log \log n)})$ where δ is the maximum node degree of input graph. They also presented a distributed construction which produces approximation solution with performance ratio $H(\frac{\delta(\delta-1)}{2})$ where $H(k) = \sum_{i=1}^{k} \frac{1}{i}$ is the harmonic function.

Ding et al. [33] found that the (ROC1)-constrained CDS is still too large and hence they further relaxed constraint (ROC1) to (ROCα) for $\alpha \geq 1$ as follows:

(ROCα) For any pair of nodes u and v, $m_C(u,v) \leq \alpha \cdot m(u,v)$ where $m(u,v)$ denotes the number of intermediate nodes in the shortest path between u and v in input graph and $m_C(u,v)$ denotes the number of intermediate nodes in the shortest path between u and v through considered CDS C.

Ding et al. [33] showed that for any $\alpha \geq 1$, computing the minimum (ROCα)-constrained CDS is NP-hard. Through computational experiments, they also found that (ROC2)-constrained CDS has size much smaller than the (ROC1)-constrained CDS. However, it is not easy to give a good theoretically guaranteed approximation for MIN-CDS with constraint (ROCα) for $\alpha \geq 2$ in general graphs and hence left the following problem open:

Open Problem 7.1.1. *Is there a polynomial-time $O(\log n)$-approximation for MIN-CDS with constraint (ROCα) for $\alpha \geq 2$?*

Du et al. [41] proved that for any $\alpha \geq 2$ MIN-CDS with constraint (ROCα) does not have a polynomial-time $(\rho \ln \delta)$-approximation for $0 < \rho < 1$ unless $NP \subseteq DTIME(n^{O(\log \log n)})$. Du et al. [42] indicated that for $\alpha \geq 2$, MIN-CDS with constraint (ROCα) could be treated as a minimum submodular cover problem with submodular objective function [108].

While it is unlikely for MIN-CDS with constraint (ROCα) ($\alpha \geq 1$) to have a polynomial-time constant-approximation in general graphs, Du et al. [43] move their attention to unit disk graphs. They constructed a polynomial-time constant-approximation in unit disk graphs for MIN-CDS with constraint (ROCα) for $\alpha \geq 5$. Du et al. [39] further gave a PTAS in unit disk graphs for MIN-CDS with constraint (ROCα) for $\alpha \geq 5$. The following are still open:

Open Problem 7.1.2. *Is there a polynomial-time constant-approximation in unit disk graphs for* MIN-CDS *with constraint (ROCα) for $1 \leq \alpha \leq 4$?*

Open Problem 7.1.3. *Is there a PTAS in unit disk graphs for* MIN-CDS *with constraint (ROCα) for $1 \leq \alpha \leq 4$?*

Ding et al. [29, 30] extended some of the above works to wireless networks with directional antennas.

7.2 Complexity in General Graphs

First, we give a simple and equivalent form of constraint (ROCα) for $\alpha = 0, 1, \ldots$. Let us define constraint (ROC*α) for $\alpha = 0, 1, \ldots$ as follows.

Let C be a considered CDS.
(ROC*0) For any pair of nodes u and v with $m(u,v) = 1$, every node adjacent to both u and v belongs to C.
(ROC*α) For any pair of nodes u and v with $m(u,v) = 1$, $m_C(u,v) \leq \alpha$.

The following lemma shows that (ROCα) and (ROC*α) are equivalent.

Lemma 7.2.1. *Let G be a connected graph and C a CDS of G. Then, for any $\alpha = 0, 1, \ldots$, C satisfies (ROCα) if and only if C satisfies (ROC*α).*

Proof. Since constraint (ROCα) contains constraint (ROC*α), it suffices to prove that (ROC*α) implies (ROCα).

For $\alpha = 0$, suppose $(u, w_1, w_2, \ldots, w_k, v)$ is a shortest path between u and v. Then (u, w_1, w_2) is a shortest path between u and w_2, (w_1, w_2, w_3) is a shortest path between w_1 and w_3, $\ldots (w_{k-1}, w_k, v)$ is a shortest path between w_{k-1} and v. By (ROC*α), all $w_1, w_2, \ldots, w_k$ belong to C.

For $\alpha \geq 1$, suppose $(u, w_1, w_2, \ldots, w_k, v)$ is a shortest path between u and v. By (ROC*α), u and w_2 are connected by a path with all intermediate nodes $s_{11}, \ldots, s_{1\alpha_1}$ in C where $\alpha_1 \leq \alpha$; $s_{1\alpha_1}$ and w_3 are connected by a path with all intermediate nodes $s_{21}, \ldots, s_{2\alpha_2}$ in C where $\alpha_2 \leq \alpha, \ldots, s_{1\alpha_{k-1}}$ and v are connected by a path with all intermediate nodes $s_{k1}, \ldots, s_{k\alpha_k}$ in C where $\alpha_k \leq \alpha$ (Fig. 7.2). Therefore, u and v are connected by a path with all intermediate nodes $s_{11}, \ldots, s_{1\alpha_1}, s_{21}, \ldots, s_{2\alpha_2}, \ldots, s_{k1}, \ldots, s_{k\alpha_k}$ in C. Hence $m_C(u,v) \leq \alpha \cdot m(u,v)$, that is, (ROC$\alpha$) holds. □

Lemma 7.2.2. *Let C be a dominating set of a connected graph G. Suppose C satisfies constraint (ROC*α) for some $\alpha \in \{0, 1, \ldots\}$. Then C is a CDS.*

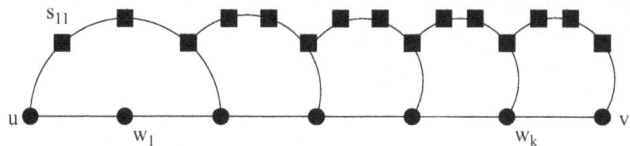

Fig. 7.2 The proof of Lemma 7.2.1

Proof. Since G is connected, we have $m(u,v) < \infty$ for any two nodes u and v. Therefore, for any two nodes $u, v \in C$, $m_C(u,v) \leq \alpha m(u,v) < \infty$, that is, there is a path connecting u and v within C. Thus, C induces a connected subgraph and hence C is a CDS. □

Now, we study the complexity of MIN-CDS with constraint (ROC*α) for $\alpha \geq 0$.

Theorem 7.2.3 (Ding et al. [31]). MIN-CDS *with constraint (ROC*0) is polynomialtime solvable.*

Proof. Let C^* be the minimum (ROC*0)-constrained CDS. Then a node u belongs to C^* if and only if u has two adjacent nodes x and y such that no edge exists between x and y. This means that for each node u, checking whether u belongs to C^* or not can be done in polynomial time. □

We show the NP-hardness of MIN-CDS with constraint (ROC*α) for $\alpha \geq 1$ by constructing reduction from the well-known NP-hard problem MIN-SET-COVER.

Theorem 7.2.4 (Ding et al. [32]). MIN-CDS *with constraint (ROC*1) is NP-hard. Moreover, it does not have polynomial-time approximation with performance ratio $\rho \ln \delta$ for $0 < \rho < 1$ unless $NP \subseteq DTIME(n^{O(\log \log n)})$, where δ is the maximum node degree of input graph.*

Proof. We construct a new reduction from MIN-SET-COVER to MIN-CDS with constraint (ROC1) as follows.

For each subset $A \in \mathcal{C}$, we create a node u_A and for each element $x \in X$, we create a node v_x. In addition, we create two nodes p and q. Connect p to every u_A for $A \in \mathcal{C}$ and connect q to every u_A for $A \in \mathcal{C}$ and every v_x for $x \in X$. Connect u_A to v_x if and only if $x \in A$. Denote by G the resulting graph (Fig. 7.3). We claim that $\mathcal{C}$ has a set cover of size at most k if and only if G has an (ROC1)-constrained CDS of size at most $k+1$.

Our claim holds trivially in case of $|\mathcal{C}| = 1$. We next assume $|\mathcal{C}| \geq 2$.

First, assume $\mathcal{C}$ has a set cover $\mathcal{A}$ of size at most k. Then, it is easy to verify that $\{u_A \mid A \in \mathcal{A}\} \cup \{q\}$ is a (ROC1)-constrained CDS of size at most $k+1$. Indeed, for a node pair $\{x,y\}$ in G, if $\{x,y\} \cap \{p,q\} = \emptyset$, then (x,q,y) must be a shortest path. If $x = p$, then we must have $y = v_z$ or $y = q$ for some $z \in X$. If $y = v_z$, then there exists $A \in \mathcal{A}$ such that $z \in A$. It follows that (p, u_A, v_z) is a shortest path. If $y = q$, then for any $A \in \mathcal{A}$, (p, u_A, q) is a shortest path.

7.2 Complexity in General Graphs 123

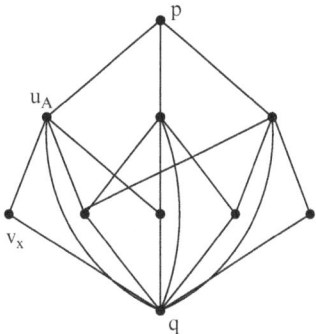

Fig. 7.3 The proof of Theorem 7.2.4

Conversely, suppose that G has an (ROC1)-constrained CDS D of size at most $k+1$. Note that the distance from p to each v_x is two and every shortest path from p to v_x for $x \in X$ must pass a node u_A for some $A \in \mathcal{C}$. Therefore, $\mathcal{A} = \{A \in \mathcal{C} \mid u_A \in D\}$ is a set cover. For $A, B \in \mathcal{C}$ with $A \neq B$, the distance between u_A and u_B is two and for every shortest path between u_A and u_B, the intermediate node is not in $\{u_A \mid A \in \mathcal{C}\}$. This means that there exists a node in D, but not in $\{u_A \mid A \in \mathcal{A}\}$. Hence, $|\mathcal{A}| \leq k$.
□

Theorem 7.2.5 (Du et al. [39]). *For $\alpha \geq 2$, MIN-CDS with constraint (ROC*α) is NP-hard. Moreover, It does not have polynomial-time approximation with performance ratio $\rho \ln \delta$ for $0 < \rho < 1$ unless $NP \subseteq DTIME(n^{O(\log \log n)})$, where δ is the maximum node degree of input graph.*

Proof. We reduce MIN-SET-COVER to MIN-CDS with constraint (ROCα) in the following way.

Consider an instance of MIN-SET-COVER, a finite set X and a collection $\mathcal{C}$ of subsets of X with $\cup_{A \in \mathcal{C}} A = X$ and $|X| \geq 2$. First, construct a bipartite graph $G = (U, V, E)$ where $U = \{u_A \mid A \in \mathcal{C}\}$, $V = \{v_x \mid x \in X\}$, and $E = \{(u_A, v_x) \mid x \in A\}$. Next, we connect all nodes in U into a complete subgraph. The obtained graph is called H (Fig. 7.4).

Now, we show that $\mathcal{C}$ contains a set cover of size at most k if and only if H has a (ROCα)-constrained CDS of size at most k where $\alpha \geq 2$.

First, assume $\mathcal{C}$ contains a set cover $\mathcal{A}$ of size at most k. Let $U_{\mathcal{A}} = \{u_A \mid A \in \mathcal{A}\}$. Clearly, $U_{\mathcal{A}}$ is a CDS. To verify $U_{\mathcal{A}}$ satisfying constraint (ROCα), we consider a pair of nodes x and y with $m(x, y) = 1$. There are two cases.

Case 1. x and y both belong to V, that is, $x = v_a$ and $y = v_b$ for some $a, b \in X$. Since $\mathcal{A}$ is a set cover. There exist $A, B \in \mathcal{A}$ such that $a \in A$ and $b \in B$. If $A \neq B$, (v_a, u_A, u_B, v_b) is a path with two intermediate nodes. If $A = B$, $(v_a, u_A = u_B, v_b)$ is a path with one intermediate node. Thus, $m_{U_{\mathcal{A}}}(x, y) \leq 2$.

Case 2. x is in U and y is in V, that is, $x = u_A$ and $y = v_b$ for some $A \in \mathcal{C}$ and $y = v_b$ with $b \notin A$. Since $\mathcal{A}$ is a set cover, there exists $B \in \mathcal{A}$ such that $b \in B$. Then (u_A, u_B, v_b) is a path with one intermediate node.

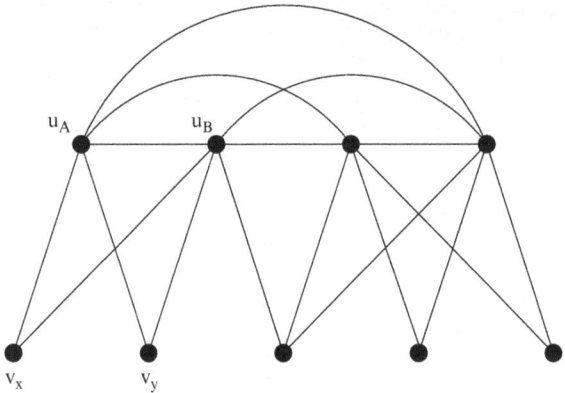

Fig. 7.4 The proof of Theorem 7.2.5

Next, assume H has a (ROCα)-constrained CDS D of size at most k. We claim that for each $v_x \in V$, we must have a node $u_A \in D$ such that $x \in A$. In fact, if $v_x \notin D$, then we must have a node u_A dominating v_x and such a u_A satisfies $x \in A$. If $v_x \in D$, then we must have another node w in D since v_x cannot dominate v_y for $y \in X - \{x\}$. To connect v_x with w, there must exist $u_A \in D$ such that $x \in A$. Our claim implies that $\{A \mid u_A \in D\}$ is a set cover.

Now, the theorem results immediately from the fact that MIN-SET-COVER has no polynomial-time $\rho \ln n$-approximation for $0 < \rho < 1$ unless $NP \subseteq DTIME(n^{O(\log \log n)})$. □

7.3 CDS with Constraint (ROC1)

Ding et al. [32] constructed two approximation algorithms for MIN-CDS with constraint (ROC1), a sequential one and a distributed one. The sequence comes from the observation that MIN-CDS with constraint (ROC1) can be reduced to MIN-SET-COVER.

Theorem 7.3.1 (Ding et al. [32]). *There is a polynomial-time* $H(\frac{\delta(\delta-1)}{2})$-*approximation for* MIN-CDS *with constraint (ROC1)*.

Proof. Let $G = (V, E)$ be an input connected graph for MIN-CDS with constraint (ROC1). To construct an instance of MIN-SET-COVER, define that

$$Q = \{\{x, y\} \mid m(x, y) = 1\}$$

and for any $w \in V$,

$$Q(w) = \{\{x, y\} \in Q \mid (x, w), (w, y) \in V\}.$$

Then, MIN-CDS with constraint (ROC1) is equivalent to the problem of finding a minimum node subset C such that $\cup_{w \in C} Q(w) = Q$, that is, MIN-SET-COVER on an input consisting of a finite set Q and a collection of subsets $\{Q(w) \mid w \in V\}$ of Q. Since MIN-SET-COVER has a polynomial-time $H(\gamma)$-approximation where γ is the maximum cardinality of a subset in input subset collection [23] and $|Q| \leq \frac{\delta(\delta-1)}{2}$, the theorem holds. □

7.4 CDS with Constraint (ROCα) for $\alpha \geq 5$

While no good approximation algorithm has been designed for MIN-CDS with constraint (ROCα) for $\alpha \geq 2$ in general graphs, progress has been made for this problem in unit disk graphs. In unit disk graphs, the size of any maximal independent set is linearly bounded by the size of the minimum CDS. Often one can use this fact to construct polynomial-time constant-approximations for various variations of the CDS problem. Du et al. [43] confirmed that this is also true for MIN-CDS with constraint (ROCα) for $\alpha \geq 5$. To present their result, let us first show a relationship between the constraint (ROC*α) for $\alpha \geq 5$ and a condition about maximal independent set as follows.

Lemma 7.4.1. *Let G be a connected graph and I a maximal independent set of G. Suppose $C \supseteq I$ is a CDS satisfies the following condition:*

(I) For any two nodes u and v in I with $m(u,v) \leq 3$, $m_C(u,v) \leq 3$.

Then constraint (ROC5) holds.

Proof. Consider any two nodes x and y with $m(x,y) = 1$. Since I is a maximal independent set, we can find nodes $x', y' \in I$ such that x' is adjacent to x and y' is adjacent to y. Therefore, $m(x',y') \leq 2 + m(x,y) = 3$. By assumption, we have $m_C(x',y') \leq 3$. Note that $I \subseteq C$. Thus, $x', y' \in C$. It follows that $m_C(x,y) \leq 5$, that is, (ROC*5) holds. By Lemma 7.2.1, (ROC5) holds. □

The following algorithm is designed based on Lemma 7.4.1.

Algorithm ROC-CDS.
  input a unit disk graph G.
  Initially, set $D \leftarrow \emptyset$. The main body of this algorithm consists of two stages:
  Stage 1. Construct a maximal independent set I.
  Stage 2. For every pair of nodes u, v in I with $m(u,v) \leq 3$, compute a shortest path $p(u,v)$ and put all intermediate nodes of $p(u,v)$ into D.
  output $C = D \cup I$.

Lemma 7.4.2. *Let I be a maximal independent set. Then for every $u \in I$, $|\{v \in I \mid 0 < m(u,v) \leq 3\}| \leq 69$.*

Proof. For each node $v \in I$ with $m(u,v) \leq 3$, v lies in the disk with center u and radius 4. By Zassenhaus–Groemer–Oler Inequality,

$$|\{v \in I \mid 0 < m(u,v) \leq 3\}| \leq \frac{2}{\sqrt{3}} \cdot 16\pi + 4\pi = 69.6\ldots$$

since $u \in I$ is not counted. Hence

$$|\{v \in I \mid 0 < m(u,v) \leq 3\}| \leq 69.$$
□

Lemma 7.4.3. *In Algorithm ROC-CDS, the node subset D obtained at Stage 2 has size $|D| \leq 207|I|/2$ where I is the maximal independent set obtained in Stage 1.*

Proof. Construct a graph H with node set I and edge set $\{(u,v) \mid u,v, \in I, 0 < m(u,v) \leq 3\}$. By Lemma 7.4.2, the maximum node degree of H is at most 69. Therefore, H contains at most $69|I|/2$ edges. In *Stage 2*, each path $p(u,v)$ corresponds an edge (u,v) of H, for which we add at most three nodes to D since $m(u,v) \leq 3$. Therefore, $|D| \leq 3 \cdot 69|I|/2 = 207|I|/2$. □

Theorem 7.4.4 (Du et al. [43]). *Algorithm ROC-CDS produces a CDS C with size*

$$|C| \leq 443\frac{2}{3} \cdot \text{opt}_{\text{MCDS}} + 201\frac{2}{3}$$

which has the property that for any pair of nodes u,v, $m_C(u,v) \leq 5m(u,v)$.

Proof. By Lemmas 7.4.3 and 3.5.6,

$$|C| \leq |D| + |I| \leq 104.5 \cdot |I| \leq 358.53 \cdot \text{opt}_{\text{MCDS}} + 503.48.$$

By Lemma 7.4.1, C has the property that for any pair of nodes u,v, $m_C(u,v) \leq 5m(u,v)$. □

Furthermore, Du et al. [41] constructed a PTAS for MIN-CDS with constraint (ROCα) for $\alpha \geq 5$.

To start their construction, let us first put input unit disk graph $G = (V,E)$ in the interior of the square $[0,q] \times [0,q]$. Then construct a grid $P(0)$ as shown in Fig. 7.5. $P(0)$ divides the square $[0,pa] \times [0,pa]$ into p^2 cells where $a = 2(\alpha+2)k$ for a positive integer k and $p = 1 + \lceil q/a \rceil$. Each cell e is a $a \times a$ square, including its left side and its lower side but not including its right side and upper side, so that all cells are disjoint and their union covers the interior of the square $[0,q] \times [0,q]$.

For each cell e, construct a $(a+4) \times (a+4)$ square and a $(a+2\alpha+4) \times (a+2\alpha+4)$ square with the center the same as that of e (Fig. 7.6). The area inside of the second square (not including the boundary of the second square) and outside of the cell e including the boundary of e) is called the *boundary area* of cell e,

7.4 CDS with Constraint (ROCα) for $\alpha \geq 5$

Fig. 7.5 Grid $P(0)$

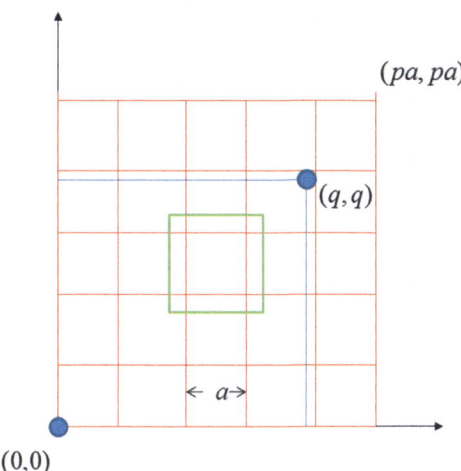

Fig. 7.6 Central area e^c and boundary area e^b

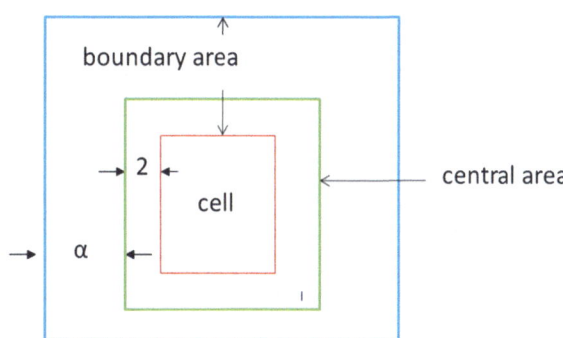

denoted by e^b. The closed area bounded by the first square is called the *central area* of cell e, denoted by e^c. The union of the boundary area and the central area is the open area bounded by the second square, denoted by e^{cb}.

Now, for each cell e, we study the following problem.

> LOCAL(e): Find the minimum subset D of nodes in $V \cap e^{cb}$ such that (a) D dominates all nodes in $V \cap e^c$, and (b) for any two nodes $u, v \in V \cap e^c$ with $m(u,v) = 1$ and $\{u,v\} \cap e \neq \emptyset$, $m_D(u,v) \leq \alpha$.

Lemma 7.4.5. *Suppose $\alpha \geq 5$ and $|V \cap e^{cb}| = n_e$. Then the minimum solution of the* LOCAL(e) *problem can be computed in time $n_e^{O(a^4)}$.*

Proof. Partition e^c into $\lceil (a+4)\sqrt{2} \rceil^2$ small squares with diameter at most one (Fig. 7.7). For each (closed) small square s, if $V \cap s \neq \emptyset$, then choose one node which would dominate all nodes in $V \cap s$. Those chosen nodes form a set D dominating $V \cap e^c$ and $|D| \leq \lceil (a+4)\sqrt{2} \rceil^2$.

Fig. 7.7 Decomposition of central area e^c

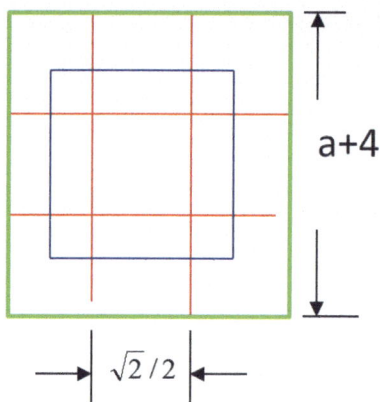

Denote by $M(u,v)$ the set of all intermediate nodes on a shortest path between u and v. For any two nodes $u, v \in D$ with $m(u,v) \leq 3$, connect them with a shortest path between u and v. Namely, set

$$C = D \cup \left(\cup_{u,v \in D: m(u,v) \leq 3} M(u,v) \right).$$

and show that C is a feasible solution of LOCAL(e) problem.

For any two nodes $u, v \in V \cap e^c$ with $m(u,v) = 1$ and $\{u, v\} \neq \emptyset$, since D dominates $V \cap e^c$, there are $u', v' \in D$ such that u is adjacent to u' and v is adjacent to v'. Thus, $m(u', v') \leq 3$. This implies that $M(u,v) \subseteq C$ and hence $m_C(u,v) \leq 5$. Therefore, C is a feasible solution of MIN-CDS with constraint (ROCα) for $\alpha \geq 5$. Moreover,

$$|C| \leq |D| + 3 \cdot \frac{|D|(|D|-1)}{2} \leq 1.5|D|^2 \leq 1.5 \cdot \lceil (a+4)\sqrt{2} \rceil^4.$$

This means that the minimum solution of the LOCAL(e) problem has a size at most $1.5 \cdot \lceil (a+4)\sqrt{2} \rceil^4$. Therefore, by an exhausting search, we can compute the minimum solution of the LOCAL(e) problem in time $n_e^{O(a^4)}$. □

Let D_e denote the minimum solution for the LOCAL(e) problem. Define $D(0) = \cup_{e \in P(0)} D_e$ where $e \in P(0)$ means that e is over all cells in partition $P(0)$.

Lemma 7.4.6. *$D(0)$ is a feasible solution of* MIN-CDS *with constraint (ROCα) for $\alpha \geq 5$ and $D(0)$ can be computed in time $n^{O(a^4)}$ where $n = |V|$.*

Proof. Since every node in V belongs to some e^c, $D(0)$ is a dominating set. Moreover, for every two nodes $u, v \in V$ with $m(u,v) = 1$, u must belong to some cell e, which implies that $u, v \in e^c$. Hence, $m_{D_e}(u,v) \leq \alpha$. It follows that $m_{D(0)}(u,v) \leq \alpha$. By Lemma 7.2.2, $D(0)$ is feasible for MIN-CDS with constraint (ROCα).

7.4 CDS with Constraint (ROCα) for $\alpha \geq 5$

Note that each node may appear in e^{cb} for at most four cells e. By Lemma 7.4.5, $D(0)$ can be computed in time

$$\sum_{e \in P(0)} n_e^{O(a^4)} \leq (4n)^{O(a^4)} = n^{O(a^4)},$$

where $n = |V|$. □

To estimate the size of $D(0)$, we consider a minimum solution D^* of MIN-CDS with constraint (ROCα). Define $P(0)^b = \cup_{e \in P(0)} e^b$.

Lemma 7.4.7. $|D(0)| \leq |D^*| + 4|D^* \cap P(0)^b|$.

Proof. First, we show that $D^* \cap e^{cb}$ is feasible for the LOCAL(e) problem. In fact, it is clear that $D^* \cap e^{cb}$ dominates $V \cap e^c$. For any two nodes $u, v \in e^c$ with $m(u,v) = 1$ and $\{u,v\} \cap e \neq \emptyset$, the path between u and v with at most α intermediate nodes must lie inside of e^{cb}. Hence, $m_{D^*}(u,v) \leq \alpha$ implies $m_{D^* \cap e^{cb}}(u,v) \leq \alpha$.

Since D_e is the optimal solution for the LOCAL(e) problem, we have $|D_e| \leq |D^* \cap e^{cb}|$. Thus

$$|D(0)| \leq \sum_{e \in P(0)} |D_e|$$

$$\leq \sum_{e \in P(0)} |D^* \cap e^{cb}|$$

$$\leq \sum_{e \in P(0)} |D^* \cap e| + \sum_{e \in P(0)} |D^* \cap e^b|$$

$$\leq |D^*| + 4|D^* \cap P(0)^b|.$$

□

Now, we shift partition $P(0)$ to $P(i)$ as shown in Fig. 7.8 such that the left and lower corner of the grid is moved to point $(-2(\alpha+2)i, -2(\alpha+2)i)$. For each $P(i)$, compute a feasible solution $D(i)$ in the same way as compute $D(0)$ for $P(0)$. Then we have

(a) $D(i)$ is a feasible solution of MIN-CDS with constraint (ROCα).
(b) $D(i)$ can be computed in time $n^{O(a^4)}$.
(c) $|D(i)| \leq |D^*| + 4|D^* \cap P(i)^b|$.

In addition, we have

Lemma 7.4.8. $|D(0)| + |D(1)| + \cdots + |D(k-1)| \leq (k+8)|D^*|$.

Proof. Note that $P(i)^b$ consists of a group of horizontal strips and a group of vertical strips (Fig. 7.9). All horizontal strips in $P(0)^b \cup P(1)^b \cup \cdots \cup P(k-1)^b$ are disjoint and all vertical strips in $P(0)^b \cup P(1)^b \cup \cdots \cup P(k-1)^b$ are also disjoint.

Fig. 7.8 Horizontal and vertical strips

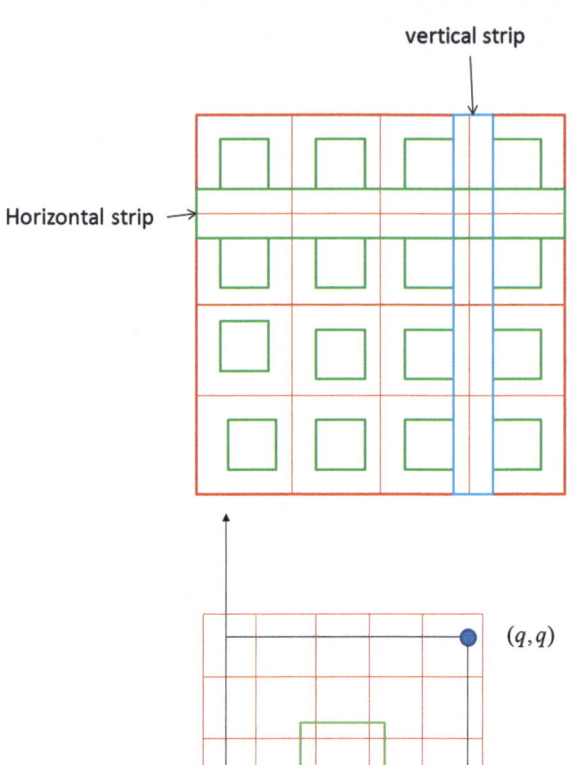

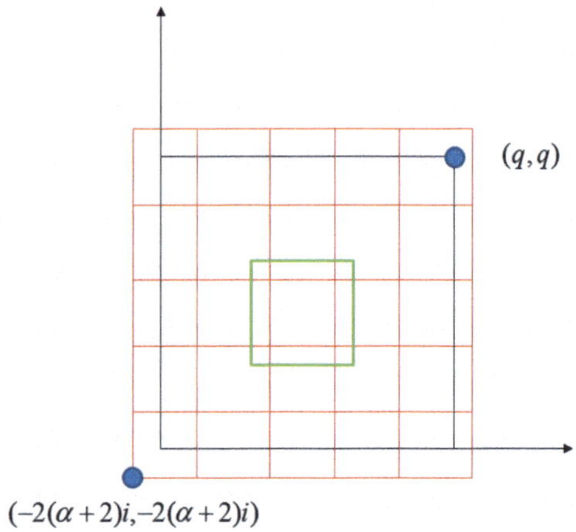

Fig. 7.9 Grid $P(i)$

Therefore,
$$\sum_{i=0}^{k-1} |D^* \cap P(i)^b| \leq 2|D^*|.$$

Hence,
$$\sum_{i=0}^{k-1} |D(i)| \leq (k+8)|D^*|. \qquad \square$$

Set $k = \lceil 1/(8\varepsilon) \rceil$ and run the following algorithm.

Algorithm PTAS
input a unit disk graph G.
  Compute $D(0), D(1), \ldots, D(k-1)$;
  Choose i^*, $0 \leq i^* \leq k-1$ such that
    $|D(i^*)| = \min(|D(0)|, |D(1)|, \ldots, |D(k-1)|)$;
output $D(i^*)$.

7.4 CDS with Constraint (ROCα) for $\alpha \geq 5$

Theorem 7.4.9 (Du et al. [41]). *Algorithm PTAS produces an approximation solution for* MIN-CDS *with constant (ROCα) with size*

$$|D(i^*)| \leq (1+\varepsilon)|D^*|$$

and running time $n^{O(1/\varepsilon^4)}$.

Proof. It follows from Lemmas 7.4.6 and 7.4.8. □

Chapter 8
CDS in Disk-Containment Graphs

> *What I'm doing here is having more*
> *points of analysis near the customers,*
> *because the key here is quick containment.*
> PETER WATKINS

8.1 Motivation and Overview

Disk-containment graphs are generalizations of the unit disk graphs. Consider a finite planar set V of nodes. Each node v is associated with a disk of radius r_v centered at v. The disk-containment graph (DCG) of V is the undirected graph $G = (V,E)$ in which $uv \in E$ if and only if the disk-associated u contains v and disk-associated v contains u. In other words, $uv \in E$ if and only if the Euclidean distance between u and v is no more than $\min\{r_u, r_v\}$. When all the disks associated with the nodes in V have unit radius, then the DCG of V is exactly the UDG of V. The DCG arises naturally from communication topologies of multihop wireless networks with disparate communication ranges [102, 124]. Indeed, if V represents the set of nodes in a multihop wireless network and each r_v represents the communication radius of the node v, the DCG of V is exactly the symmetric communication topology of the multihop wireless network.

In this chapter, we present the design and analysis of an approximation algorithm for MIN-CDS in a DCG G [112]. By proper scaling, we assume that the smallest radius of the associated disks is one and the largest radius radius of the associated disks is R. Let $g = \frac{1+\sqrt{5}}{2}$ be the golden ratio, and denote

$$R^* = 5 + 8\lceil \log_g R \rceil.$$

The *local independence number* of a node u is the largest size of the independent sets contained in the closed neighborhood of u in G. We first show that the local independence number of any node in G is at most R^* in Sect. 8.2. Based on this upper bound, we derive in Sect. 8.3 a relation between the independence number α (the size of a maximum independent set) of G and connected domination number γ_c (the size of a minimum connected dominating set) of G:

$$\alpha \leq (R^* - 1)\gamma_c + 1.$$

The relation between the independence number α and the connected domination number γ_c plays a key role in deriving the approximation bounds of various two-phased greedy approximation algorithms adapted for MIN-CDS of multihop wireless networks with disparate communication ranges [102, 106, 124]. In this chapter, we first proved that $\alpha \leq (R^* - 1)\gamma_c + 1$, where $R^* = 5 + 8\lceil \log_g R \rceil$ for any $R \geq 1$. From this relation, we then derived an approximation bound $R^* + \ln(R^* - 2) + 1$ of the two-phased greedy approximation algorithm adapted from [106].

Tighter relation between α and γ_c may be derived with more sophisticated analyses. A possible approach of obtaining tighter relation between α and γ_c is to develop a tighter bound on the number of independent nodes that can be packed in the neighborhood of a pair of adjacent nodes. An attempt along this approach has been made in [124], but the argument in [124] contains a critical error. However, we do believe that this approach is very promising to achieve tighter relation between α and γ_c.

Throughout this chapter, $D(u,r)$ denotes the *closed* disk of radius r centered at u, and $\partial D(u,r)$ denotes the boundary circle of $D(u,r)$. The Euclidean distance between two nodes u and v is denoted by $\|uv\|$. The cardinality of a finite set S is denoted by $|S|$.

8.2 Local Independence Number

In this section, we present an upper bound on the local independence number of an arbitrary node u.

Theorem 8.2.1. *Suppose that I is an independent set of nodes adjacent to a node u. Then $|I| \leq R^*$.*

The rest of this section is devoted to the proof of Theorem 8.2.1. Consider an arbitrary node $u \in V$ and an independent set I of nodes adjacent to a node u. Let I_1 be the set of nodes in I lying in the closed disk of radius g centered at u, and for each $j \geq 2$ let

$$I_j = \{v \in I : g^{j-1} < \|uv\| \leq g^j\}.$$

From [49] we have $|I_1| \leq 12$. We shall further prove the following lemma for $j \geq 2$.

Lemma 8.2.2. *For any $j \geq 2$, $|I_j| \leq 9$ and $|I_j \cup I_{j+1}| \leq 16$.*

The above lemma implies Theorem 8.2.1 immediately. If $\lceil \log_g R \rceil$ is odd, then

8.2 Local Independence Number

$$|I| = \left| \bigcup_{j=1}^{\lceil \log_g R \rceil} I_j \right|$$

$$= |I_1| + \sum_{i=1}^{(\lceil \log_g R \rceil - 1)/2} |I_{2i} \cup I_{2i+1}|$$

$$\leq 12 + 16 \cdot \left(\lceil \log_g R \rceil - 1\right)/2$$

$$= 8 \lceil \log_g R \rceil + 4 < R^*.$$

If $\lceil \log_g R \rceil$ is even, then

$$|I| = \left| \bigcup_{j=1}^{\lceil \log_g R \rceil} I_j \right|$$

$$= |I_1| + |I_2| + \sum_{i=2}^{\lceil \log_g R \rceil/2 - 1} |I_{2i-1} \cup I_{2i}|$$

$$\leq 12 + 9 + 16 \left(\lceil \log_g R \rceil/2 - 1\right)$$

$$= 8 \lceil \log_g R \rceil + 5 = R^*.$$

So, Theorem 8.2.1 holds in either case.

Next, we prove Lemma 8.2.2 by using a subtle angular argument. Fix a $j \geq 2$. We first derive a lower bound on the angle separation between any pair of nodes in I_j at u.

Lemma 8.2.3. *Suppose that v and w are two distinct nodes in I_j satisfying that $\|uv\| \geq \|uw\|$. Then, $\angle wuv > 36°$. In addition, for any $36° \leq \alpha < 60°$,*

1. *If $\|uw\| \geq 2g^{j-1} \cos \alpha$, then $\angle wuv > \arccos \frac{g}{4\cos \alpha}$;*
2. *If $\|uv\| \leq 2g^{j-1} \cos \alpha$, then $\angle wuv > \alpha$.*

Proof. Since v and w are two independent neighbors of u, we have

$$\|vw\| > \min\{r_v, r_w\} \geq \min\{\|uv\|, \|uw\|\} = \|uw\|.$$

Thus, v is outside the disk $D(w, \|uw\|)$. Since

$$2\|uw\| > 2g^{j-1} > g^j,$$

the two circles $\partial D(u, g^j)$ and $\partial D(w, \|uw\|)$ intersect. Let z denote their intersection point which lies on the same side of line uw as v (see Fig. 8.1). Then

$$\cos \angle wuz = \frac{\|uz\|}{2\|uw\|} < \frac{g^j}{2g^{j-1}} = \frac{g}{2} = \cos 36°,$$

Fig. 8.1 v and w are two distinct nodes in I_j satisfying that $\|uv\| \geq \|uw\|$

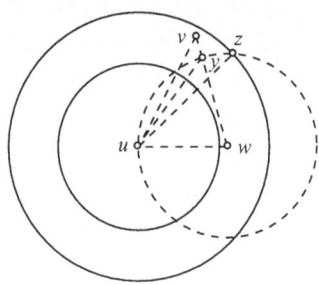

which implies $\angle wuz > 36°$. Hence,

$$\angle wuv \geq \angle wuz > 36°.$$

Clearly, $\angle wuv = 36°$ if and only if $w \in \partial D\left(u, g^{j-1}\right)$ and v is coincide with the point z.

(1) Suppose that $\|uw\| \geq 2g^{j-1}\cos\alpha$. We have

$$\cos \angle wuz = \frac{\|uz\|}{2\|uw\|} \leq \frac{g^j}{4g^{j-1}\cos\alpha} = \frac{g}{4\cos\alpha},$$

which implies $\angle wuz \geq \arccos \frac{g}{4\cos\alpha}$. Since v and w are independent, v is outside the disk $D(w, \|uw\|)$. Thus,

$$\angle wuv > \angle wuz \geq \arccos \frac{g}{4\cos\alpha}.$$

(2) Suppose that $\|uv\| \leq 2g^{j-1}\cos\alpha$. Let y be the intersection point of the line segment vw and $\partial D(w, \|uw\|)$. Then,

$$\|uy\| < \|uv\| \leq 2g^{j-1}\cos\alpha.$$

So,

$$\cos \angle wuy = \frac{\|uy\|}{2\|uw\|} < \frac{2g^{j-1}\cos\alpha}{2g^{j-1}} = \cos\alpha,$$

which implies $\angle wuy > \alpha$.

$$\angle wuv > \angle wuy > \alpha.$$

This completes the proof for lemma. □

8.2 Local Independence Number

The first inequality in Lemma 8.2.2 follows immediately from the above lemma. The next lemma derives some necessary conditions for $|I_j| = 9$.

Lemma 8.2.4. *Suppose that I_j consists of nine nodes $v_1, v_2, \ldots, v_9$ sorted in the increasing order of the distances from u. Then*

1. $\|uv_1\| \leq 2g^{j-1}\cos 58.6°$ *and* $\|uv_9\| \geq 2g^{j-1}\cos 39°$;
2. $\|uv_2\| \leq 2g^{j-1}\cos 58.2°$ *and* $\|uv_8\| \geq 2g^{j-1}\cos 39.8°$;
3. $\|uv_3\| \leq 2g^{j-1}\cos 56.29°$ *and* $\|uv_7\| \geq 2g^{j-1}\cos 43.2°$.

Proof. We will use the following fact multiple times in this proof: Suppose that I' is a subset of five nodes in I_j. Then, among five consecutive sectors centered at u formed by the five nodes in I', at least one of them does not contain any other node in I_j. This is because $|I_j \setminus I'| = 4 < 5$ and hence at least one of those five sectors does not contain any node in $I_j \setminus I'$.

(1) We prove the first part of lemma by contradiction. Assume to the contrary that either

$$\|uv_1\| > 2g^{j-1}\cos 58.6°$$

or

$$\|uv_9\| < 2g^{j-1}\cos 39°.$$

We first claim that the angle separation of any two nodes in I_j at u is greater than $39°$. Indeed, if

$$\|uv_1\| > 2g^{j-1}\cos 58.6°,$$

then

$$\|uv_i\| > 2g^{j-1}\cos 58.6°$$

for all $1 \leq i \leq 9$, and hence the claim holds by Lemma 8.2.3(1). If

$$\|uv_9\| < 2g^{j-1}\cos 39°,$$

then

$$\|uv_i\| < 2g^{j-1}\cos 39°$$

for all $1 \leq i \leq 9$, and hence the claim holds by Lemma 8.2.3(2). So, our claim is true. We proceed in two cases.

Case 1. $\|uv_5\| \geq 2g^{j-1}\cos 50°$. Let v_i and v_k be the two nodes in $\{v_5, v_6, \ldots, v_9\}$ such that the sector $\angle v_i u v_k$ centered at u does not contain any other node in I_j. Then by Lemma 8.2.3(1), $\angle v_i u v_k > 51°$. So, the total of the nine consecutive angles at u formed by the nodes in I_j is greater than

$$51° + 8 \cdot 39° = 363° > 360°,$$

which is also a contradiction.

Case 2. $\|uv_5\| < 2g^{j-1}\cos 50°$. Let v_i and v_k be the two nodes in $\{v_1, v_2, \ldots, v_5\}$ such that the sector $\angle v_i u v_k$ centered at u does not contain any other node in I_j. Then by Lemma 8.2.3(2), $\angle v_i u v_k > 50°$. So, the total of the nine consecutive angles at u formed by the nodes in I_j is greater than

$$50° + 8 \cdot 39° = 362° > 360°,$$

which is a contradiction.

In either case, we have reached a contradiction. Therefore, the first part of the lemma holds.

(2) We prove the second part of the lemma by contradiction. Assume to the contrary that either

$$\|uv_2\| > 2g^{j-1}\cos 58.2°,$$

or

$$\|uv_8\| < 2g^{j-1}\cos 39.8°.$$

We first claim that there exists a node $v_a \in I_j$ such that the angle separation of any two nodes in $I_j \setminus \{v_a\}$ at u is greater than $39.8°$. Indeed, if

$$\|uv_2\| > 2g^{j-1}\cos 58.2°,$$

then

$$\|uv_i\| > 2g^{j-1}\cos 58.2°$$

for all $2 \leq i \leq 9$, and hence the claim holds for $a = 1$ by Lemma 8.2.3(1). If

$$\|uv_8\| < 2g^{j-1}\cos 39.8°,$$

then

$$\|uv_i\| < 2g^{j-1}\cos 39.8°$$

for all $1 \leq i \leq 8$, and hence the claim holds for $a = 9$ by Lemma 8.2.3(2). So, our claim is true. We remark that the angle separation between v_a and any other node is still greater than $36°$. We proceed in two cases.

Case 1. $\|uv_5\| \geq 2g^{j-1}\cos 50°$. Let v_i and v_k be the two nodes in $\{v_5, v_6, \ldots, v_9\}$ such that the sector $\angle v_i u v_k$ centered at u does not contain any other node in I_j. Then by Lemma 8.2.3(1), $\angle v_i u v_k > 51°$. Let k be the number of consecutive angles at u formed by the nodes in I_j other than $\angle v_i u v_k$ with v_a on the boundary. Then, $k \leq 2$. So, the total of the nine consecutive angles at u formed by the nodes in I_j is greater than

$$51° + (8-k) \cdot 39.8° + k \cdot 36°$$
$$= 51° + 8 \cdot 39.8° - k \cdot 3.8°$$
$$\geq 51° + 8 \cdot 39.8° - 2 \cdot 3.8°$$
$$= 361.8° > 360°,$$

which is also a contradiction.

8.2 Local Independence Number

Case 2. $\|uv_5\| < 2g^{j-1}\cos 50°$. Let v_i and v_k be the two nodes in $\{v_1, v_2, \ldots, v_5\}$ such that the sector $\angle v_i uv_k$ centered at u does not contain any other node in I_j. Then by Lemma 8.2.3(2), $\angle v_i uv_k > 50°$. Let k be the number of consecutive angles at u formed by the nodes in I_j other than $\angle v_i uv_k$ with v_a on the boundary. Then, $k \leq 2$. So, the total of the nine consecutive angles at u formed by the nodes in I_j is greater than

$$50° + (8-k) \cdot 39.8° + k \cdot 36° = 50° + 8 \cdot 39.8° - k \cdot 3.8°$$
$$\geq 50° + 8 \cdot 39.8° - 2 \cdot 3.8°$$
$$= 360.8° > 360°,$$

which is a contradiction.

In either case, we have reached a contradiction. Therefore, the first part of the lemma holds.

(3) We prove the third part of the lemma by contradiction. Assume to the contrary that either

$$\|uv_3\| > 2g^{j-1}\cos 56.29°$$

or

$$\|uv_7\| < 2g^{j-1}\cos 43.2°.$$

We claim that there exist two nodes $v_a, v_b \in I_j$ such that $\angle v_a uv_b > 58.2°$ and the angle separation at u of any two nodes in $I' = I_j \setminus \{v_a, v_b\}$ is greater than $43.2°$. Indeed, if

$$\|uv_3\| > 2g^{j-1}\cos 56.29°,$$

then

$$\|uv_i\| > 2g^{j-1}\cos 56.29°$$

for all $3 \leq i \leq 9$ and hence the angle separation at u of any two nodes in $I_j \setminus \{v_1, v_2\}$ is greater than $43.2°$ by Lemma 8.2.3(1). By the second part of this lemma, we have

$$\|uv_2\| \leq 2g^{j-1}\cos 58.2°,$$

which implies $\angle v_1 uv_2 > 58.2°$ by Lemma 8.2.3(2). Thus the claim holds with $a = 1$ and $b = 2$. Similarly, if

$$\|uv_7\| < 2g^{j-1}\cos 43.2°,$$

then

$$\|uv_i\| < 2g^{j-1}\cos 43.2°$$

for all $1 \leq i \leq 7$ and hence the angle separation at u of any two nodes in $I_j \setminus \{v_8, v_9\}$ is greater than $43.2°$ by Lemma 8.2.3(2). By the second part of this lemma, we have

$$\|uv_8\| \geq 2g^{j-1} \cos 39.8°,$$

which implies that $\angle v_8 u v_9 > 58.2°$ by Lemma 8.2.3(1). Thus the claim holds with $a = 8$ and $b = 9$. So, our claim is true. We proceed in two cases.

Case 1. The sector $\angle v_a u v_b$ centered at u does not contain any node in I'. Then, among the nine consecutive angles at u formed by the nodes in I_j, $\angle v_a u v_b$ is greater than $58.2°$, the two other angles with v_a and v_b on the boundary respectively are each greater than $36°$, and the rest six angles are all greater than $43.2°$. So, the total of these nine angles is greater than

$$58.2° + 2 \cdot 36° + 6 \cdot 43.2° = 389.4° > 360°,$$

which is a contradiction.

Case 2. The sector $\angle v_a u v_b$ centered at u contains at least one node in I'. Then, among the nine consecutive angles at u formed by the nodes in I_j, the four angles with v_a and v_b on the boundary respectively are each greater than $36°$, and the rest five angles are all greater than $43.2°$. So, the total of these nine angles is greater than

$$4 \cdot 36° + 5 \cdot 43.2° = 360°,$$

which is also a contradiction.

In either case, we have reached a contradiction. Therefore, the first part of the lemma holds. □

We further derive a lower bound on the angle separation between any pair of nodes in I_j and I_{j+1} respectively at u.

Lemma 8.2.5. *Suppose that $w \in I_j$ and $v \in I_{j+1}$.*

1. *If $\|uw\| \geq 2g^{j-1} \cos \alpha$ for some $36° \leq \alpha \leq \arccos \frac{g^2}{4}$, then $\angle wuv > \arccos \frac{g^2}{4 \cos \alpha}$.*
2. *If $\|uv\| \leq 2g^j \cos \alpha$ for some $\arccos \frac{1}{g} \leq \alpha < 60°$, then $\angle wuv > \arccos(g \cos \alpha)$.*

Proof. Since v and w are two independent neighbors of u and $\|uv\| > \|uw\|$, we have

$$\|vw\| > \min\{r_v, r_w\} \geq \min\{\|uv\|, \|uw\|\} = \|uw\|.$$

Thus, v is outside the disk $D(w, \|uw\|)$.

(1) Since

$$\alpha \leq \arccos \frac{g^2}{4},$$

we have

$$2\|uw\| \geq 4g^{j-1} \cos \alpha \geq 4g^{j-1} \frac{g^2}{4} = g^{j+1}.$$

8.2 Local Independence Number

Fig. 8.2 Part (1) in the proof of Lemma 8.2.5

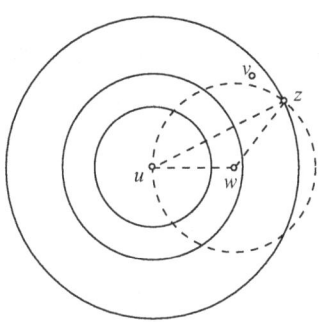

Thus, the two circles $\partial D(u, g^{j+1})$ and $\partial D(w, \|uw\|)$ intersect. Let z denote their intersection point which lies on the same side of line uw as v (see Fig. 8.2). Since

$$\|uw\| \geq 2g^{j-1} \cos \alpha,$$

we have

$$\cos \angle wuz = \frac{\|uz\|}{2\|uw\|} \leq \frac{g^{j+1}}{4g^{j-1} \cos \alpha} = \frac{g^2}{4 \cos \alpha},$$

which implies that

$$\angle wuz \geq \arccos \frac{g^2}{4 \cos \alpha}.$$

Thus,

$$\angle wuv > \angle wuz \geq \arccos \frac{g^2}{4 \cos \alpha}.$$

(2) Since

$$\alpha \geq \arccos \frac{1}{g},$$

we have

$$\|uv\| \leq 2g^j \cos \alpha \leq 2g^j \frac{1}{g} = 2g^{j-1} < 2\|uw\|.$$

Thus, the two circles $\partial D(u, \|uv\|)$ and $\partial D(w, \|uw\|)$ intersect. Let y denote their intersection point which lies on the same side of line uw as v (see Fig. 8.3). Since

$$\|uv\| \leq 2g^j \cos \alpha,$$

we have

$$\cos \angle wuy = \frac{\|yu\|}{2\|wu\|} < \frac{2g^j \cos \alpha}{2g^{j-1}} = g \cos \alpha,$$

which implies that

$$\angle wuy > \arccos(g \cos \alpha).$$

Fig. 8.3 Part (2) in the proof of Lemma 8.2.5

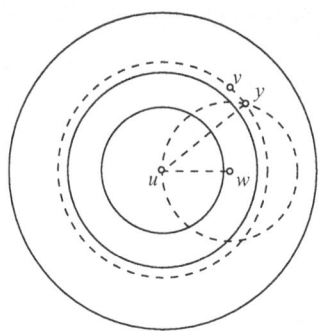

Thus,

$$\angle wuv > \angle wuy > \arccos(g\cos\alpha).$$

This completes the proof for lemma. □

Now are ready to prove the second inequality in Lemma 8.2.2. Assume to the contrary that $|I_j \cup I_{j+1}| = l \geq 17$. Let

$$I_j \cup I_{j+1} = \{v_i : 1 \leq i \leq l\}$$

where $v_1, v_2, \ldots, v_l$ are sorted in the increasing order of the distances from the node u. By Lemma 8.2.2, we have

$$\max\{|I_j|, |I_{j+1}|\} \leq 9.$$

Since $l \geq 17$, we must have

$$\max\{|I_j|, |I_{j+1}|\} = 9,$$
$$\min\{|I_j|, |I_{j+1}|\} \geq 8.$$

We consider two cases:

Case 1. $|I_j| = 9$. Then $|I_{j+1}| \geq 8$. By Lemma 8.2.4, we have

$$\|uv_7\| \geq 2g^{j-1}\cos 43.2°.$$

Let $J = \{v_7, v_8, v_9\}$. By Lemma 8.2.3(1), the angle separation between any two nodes in J at u is greater than $56.29°$. We further consider two subcases:

Subcase 1.1: There exist two nodes $v_a, v_b \in J$ such that the sector $\angle v_a u v_b$ centered at u does not contain any node in I_{j+1} (see Fig. 8.4). Let v_i and v_k be the two nodes in I_{j+1} such that the sector $\angle v_i u v_k$ contains v_a and v_b but does not contain any other node in I_{j+1}, and v_i, v_a, v_b and v_k are in the clockwise direction with respect to u. By Lemma 8.2.5(1),

$$\min\{\angle v_k u v_b, \angle v_a u v_i\} > 26°.$$

8.2 Local Independence Number

Fig. 8.4 Subcase 1.1

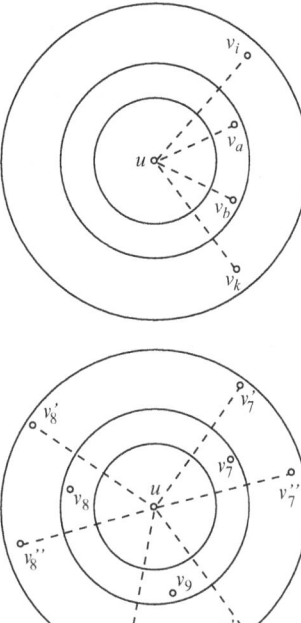

Fig. 8.5 Subcase 1.2

Thus,

$$\angle v_k u v_b + \angle v_b u v_a + \angle v_a u v_i$$
$$> 2 \cdot 26° + 56.29°$$
$$= 108.29°.$$

Hence, the total of the $|I_{j+1}|$ consecutive angles at u formed by the nodes in I_{j+1} is greater than

$$108.29° + (|I_{j+1}| - 1) \cdot 36°$$
$$\geq 108.29° + 7 \cdot 36° = 360.29°$$
$$> 360°,$$

which is a contradiction.

Subcase 1.2: For any two nodes $v_a, v_b \in J$, the sector $\angle v_a u v_b$ centered at u contains at least one node in I_{j+1} (see Fig. 8.5). For each $a = 7, 8$ and 9, let $v'_a, v''_a \in I_{j+1}$ satisfy that v_a is the only node contained in the sector $\angle v'_a u v''_a$ centered at u among all the nodes in $I_{j+1} \cup J$. Then by Lemmas 8.2.5(1) and 8.2.4, we have

Fig. 8.6 Subcase 2.1

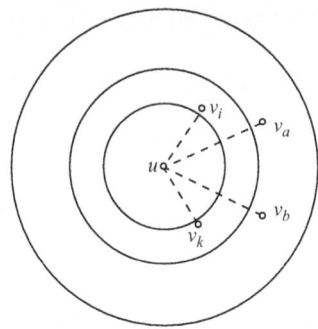

$$\min\{\angle v'_7 uv_7, \angle v_7 uv''_7\} > 26°,$$
$$\min\{\angle v'_8 uv_8, \angle v_8 uv''_8\} > 31.5°,$$
$$\min\{\angle v'_9 uv_9, \angle v_9 uv''_9\} > 32.5°.$$

Thus,

$$\angle v'_7 uv''_7 + \angle v'_8 uv''_8 + \angle v'_9 uv''_9$$
$$> 2 \cdot (26° + 31.5° + 32.5°)$$
$$= 180°.$$

Hence, the total of the $|I_{j+1}|$ consecutive angles at u formed by the nodes in I_{j+1} is greater than

$$180° + (|I_{j+1}| - 3) \cdot 36°$$
$$\geq 180° + 5 \cdot 36° = 360°,$$

which is a contradiction.

Case 2. $|I_j| = 8$. Then $|I_{j+1}| = 9$. By Lemma 8.2.4, we have

$$\|uv_{11}\| \leq 2g^j \cos 56.29°.$$

Let $J = \{v_9, v_{10}, v_{11}\}$. By Lemma 8.2.3(2), the angle separation between any two nodes in J at u is greater than $56.29°$. We further consider two subcases:

Subcase 2.1: There exist two nodes $v_a, v_b \in J$ such that the sector $\angle v_a u v_b$ centered at u does not contain any node in I_j (see Fig. 8.6). Let v_i and v_k be the two nodes in I_j such that the sector $\angle v_i u v_k$ contains v_a and v_b but does not contain any other node in I_j, and v_i, v_a, v_b and v_k are in the clockwise direction with respect to u. By Lemma 8.2.5(2),

$$\min\{\angle v_k u v_b, \angle v_a u v_i\} > 26°.$$

8.2 Local Independence Number

Fig. 8.7 Subcase 2.2

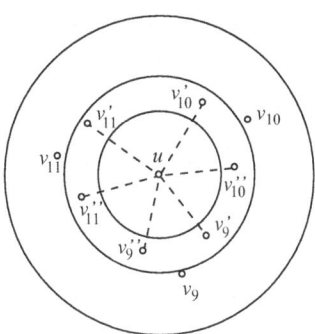

Thus,

$$\angle v_k u v_b + \angle v_b u v_a + \angle v_a u v_i$$
$$> 2 \cdot 26° + 56.29°$$
$$= 108.29°.$$

Hence, the total of the 8 consecutive angles at u formed by the nodes in I_j is greater than

$$108.29° + 7 \cdot 36° = 360.29° > 360°,$$

which is a contradiction.

Subcase 2.2: For any two nodes $v_a, v_b \in J$, the sector $\angle v_a u v_b$ centered at u contains at least one node in I_j (see Fig. 8.7). For each $a = 9, 10$ and 11, let $v'_a, v''_a \in I_j$ satisfying that v_a is the only node contained in the sector $\angle v'_a u v''_a$ centered at u among all the nodes in $I_j \cup J$. Then by Lemmas 8.2.5(2) and 8.2.4, we have

$$\min\{\angle v'_9 u v_9, \angle v_9 u v''_9\} > 32.5°,$$
$$\min\{\angle v'_{10} u v_{10}, \angle v_{10} u v''_{10}\} > 31.5°,$$
$$\min\{\angle v'_{11} u v_{11}, \angle v_{11} u v''_{11}\} > 26°.$$

Thus,

$$\angle v'_9 u v''_9 + \angle v'_{10} u v''_{10} + \angle v'_{11} u v''_{11}$$
$$> 2 \cdot (32.5° + +31.5° + 26°)$$
$$= 180°.$$

Hence, the total of the eight consecutive angles at u formed by the nodes in I_j is greater than

$$180° + 5 \cdot 36° = 360°,$$

which is a contradiction.

Thus, in every case we have reached a contradiction. So, we must have $|I_j \cup I_{j+1}| \leq 16$. This completes the proof of Lemma 8.2.2.

8.3 Independence Number

In this section, we present an upper bound on the independence number α in terms of the connected domination number γ_c.

Theorem 8.3.1. $\alpha \leq (R^* - 1)\gamma_c + 1$.

Proof. Let M be any maximum independent set of G, and OPT be any MIN-CDS of G. Then $|M| = \alpha$ and $|OPT| = \gamma_c$. Consider an arbitrary preorder traversal of $G[OPT]$ given by v_j with $1 \leq j \leq \gamma_c$. Let M_1 be the set of nodes in M that are adjacent to v_1. For any $2 \leq j \leq \gamma_c$, let M_j be the set of nodes in M that are adjacent to v_j but none of $v_1, v_2, \ldots, v_{j-1}$. Then the γ_c sets M_j with $1 \leq j \leq \gamma_c$ form a partition of M. By Theorem 8.2.1, $|M_1| \leq R^*$. For any $2 \leq j \leq \gamma_c$, there exists an index $1 \leq j' \leq j-1$ such that $v_{j'}$ is adjacent to v_j. Since $v_{j'}$ is not adjacent to any node in M_j, the set $\{v_{j'}\} \cup M_j$ is an independent set of nodes adjacent to v_j. Again by Theorem 8.2.1, we have

$$|M_j| + 1 \leq R^*,$$

and consequently

$$|M_j| \leq R^* - 1.$$

Therefore,

$$|M| = \sum_{j=1}^{\gamma_c} |M_j| \leq R^* + (R^* - 1)(\gamma_c - 1)$$
$$= (R^* - 1)\gamma_c + 1.$$

Thus, the theorem holds. □

8.4 Greedy Approximation for MIN-CDS

In this section, we present a greedy algorithm adapted from the two-phased greedy approximation algorithm originally proposed in [106] for computing a CDS in UDGs to DCGs.

The greedy algorithm consists of two phases. The first phase selects a maximal independent set (MIS) I of G. Specifically, we construct an arbitrary rooted spanning tree T of G and select an MIS I of G in the first-fit manner in the breadth-first-search ordering in T. The second phase selects a set C of connectors to interconnect I. For any subset $U \subseteq V \setminus I$, $f(U)$ denotes the number of connected components in $G[I \cup U]$. For any $U \subseteq V \setminus I$ and any $w \in V \setminus I$, the *gain* of w with respect to U is defined to be $f(U) - f(U \cup \{w\})$. The second phase greedily selects C iteratively as follows. Initially C is empty. While $f(C) > 1$, choose a node $w \in V \setminus (I \cup C)$ with *maximum* gain with respect to C and add w to C. When $f(C) = 1$, then $I \cup C$ is a CDS. Let C be the output of the second phase. Then, $I \cup C$ is the output CDS.

8.4 Greedy Approximation for MIN-CDS

The correctness of the second phase follows from the following bound on the gain established in [106].

Lemma 8.4.1. *Suppose that there are $f(U) > 1$ for some $U \subseteq V \setminus I$. Then, there exists a $w \in V \setminus (I \cup U)$ whose gain with respect to U is at least*

$$\max\{1, \lceil f(U)/\gamma_c \rceil - 1\}.$$

Proof. By the selection of I, any pair of complementary subset of I are separated by exactly two hops. Thus, there is a node w which is adjacent to at least two connected components of $G[I \cup U]$. For such node w, its gain with respect to U is at least one and $w \in V \setminus (I \cup U)$.

Now, consider a minimum CDS

$$OPT = \{v_i : 1 \leq i \leq \gamma_c\}$$

of G, and let d_i be the number of components adjacent to or containing v_i for $1 \leq i \leq \gamma_c$. Then,

$$\sum_{i=1}^{\gamma_c} d_i \geq f(U)$$

because each component of $G[I \cup U]$ must be adjacent to or contain some node in OPT. So,

$$\max_{1 \leq i \leq \gamma_c} d_i \geq \lceil f(U)/\gamma_c \rceil.$$

Let w be the node in OPT which is adjacent to the largest number of connected components in $G[I \cup U]$. Then, the gain of w with respect to U is at least $\lceil f(U)/\gamma_c \rceil - 1$ and $w \in V \setminus (I \cup U)$.

Finally, let w be a node in $w \in V \setminus (I \cup U)$ with maximum gain with respect to U. Then, its gain with respect to U is at least

$$\max\{1, \lceil f(U)/\gamma_c \rceil - 1\}.$$

So, the lemma holds. □

We further apply the above lemma to derive the following upper bound on $|C|$.

Lemma 8.4.2. $|C| \leq (\ln(R^* - 2) + 2)\gamma_c$.

Proof. For each $1 \leq i \leq |C|$, we denote by C_i the sequence of the first i nodes in C. We also set $C_0 = \emptyset$. Let k be the first (smallest) nonnegative integer such that

$$f(C_k) < 2\gamma_c + 2.$$

We claim that

$$|C \setminus C_k| \leq 2\gamma_c - 1.$$

By Lemma 8.4.1, each node in $C \setminus C_k$ has gain at least one. If $f(C_k) \leq 2\gamma_c$, then

$$|C \setminus C_k| \leq f(C_k) - 1 \leq 2\gamma_c - 1.$$

If $f(C_k) = 2\gamma_c + 1$, then the first node in $C \setminus C_k$ has gain at least two with respect to C_k by Lemma 8.4.1, and hence

$$2 + (|C \setminus C_k| - 1) \leq f(C_k) - 1 = 2\gamma_c,$$

which also implies that

$$|C \setminus C_k| \leq 2\gamma_c - 1.$$

Thus, the claim holds.

The previous claim implies that

$$|C| = k + |C \setminus C_k| \leq k + 2\gamma_c - 1 = (k-1) + 2\gamma_c.$$

Thus, it is sufficient to show that

$$k - 1 \leq \gamma_c \ln(\Delta - 2).$$

This inequality holds trivially if $k \leq 1$. So we assume that $k > 1$. For each $0 \leq i \leq k$, let

$$\ell_i = f(C_i) - \gamma_c.$$

Then

$$|I| - \gamma_c = \ell_0 > \ell_1 > \cdots > \ell_k \geq \gamma_c + 2.$$

By Lemma 8.4.1, for each $0 \leq i \leq k$,

$$\ell_{i-1} - \ell_i = f(C_{i-1}) - f(C_i) \geq \frac{f(C_{i-1})}{\gamma_c} - 1 = \frac{\ell_{i-1}}{\gamma_c},$$

and hence

$$\frac{\ell_{i-1} - \ell_i}{\ell_{i-1}} \geq \frac{1}{\gamma_c}.$$

Therefore,

$$\frac{k}{\gamma_c} \leq \sum_{i=1}^{k} \frac{\ell_{i-1} - \ell_i}{\ell_{i-1}} \leq \ln \frac{\ell_0}{\ell_k} \leq \ln \frac{|I| - \gamma_c}{\gamma_c + 2}.$$

By Theorem 8.3.1,

$$\frac{|I| - \gamma_c}{\gamma_c + 2} \leq \frac{(R^* - 1)\gamma_c + 1 - \gamma_c}{\gamma_c + 2}$$

$$= \frac{(R^* - 2)\gamma_c + 1}{\gamma_c + 2}$$

$$\leq R^* - 2.$$

8.4 Greedy Approximation for MIN-CDS

Thus,
$$\frac{k}{\gamma_c} \leq \ln(R^* - 2)$$
which implies
$$k \leq \gamma_c \ln(R^* - 2).$$

This completes the proof of the lemma. □

From Theorem 8.3.1 and Lemma 8.4.2, we obtain the following bound on the size of the CDS output by the greedy algorithm.

Theorem 8.4.3. $|I \cup C| \leq (R^* + \ln(R^* - 2) + 1)\gamma_c + 1.$

Chapter 9
CDS in Disk-Intersection Graphs

> *I don't like to hurt people, I really*
> *don't like it at all. But in order to get a red light*
> *at the intersection, you sometimes have to have an accident.*
> JACK ANDERSON

9.1 Motivation and Overview

Consider a finite set V of nodes in the plane and a radius function $r : V \to \mathbb{R}^+$. The *disk-intersection graph* (DIG) of V with the radius function r, denoted by $G_r(V)$, is the undirected graph on V in which u and v are adjacent if and only if the disk centered at u of radius $r(u)$ and the disk centered at v of radius $r(v)$ intersect, or equivalently,

$$\|uv\| \leq r(u) + r(v).$$

If $r(v) = 1/2$ for all $v \in V$, then $G_r(V)$ is exactly the unit disk graph (UDG) of V. Thus, the class of UDGs is a subclass of the class of DIGs. Hence, MIN-DS and MIN-CDS restricted to DIGs are also NP-hard. However, the approximation algorithms for MIN-DS and MIN-CDS restricted to UDGs cannot be directly extended to those for MIN-DS and MIN-CDS restricted to DIGs.

In this chapter, we present a simple local-search approximation algorithm for MIN-DS of DIGs, which yields a polynomial time approximation scheme (PTAS) for MIN-DS of DIGs [59]. In addition, we show that for any fixed $\varepsilon > 0$, there is a polynomial $(3 + \varepsilon)$-approximation algorithm for MIN-CDS of DIGs. The rest of this chapter is organized as follows. In Sect. 9.2, we introduce the Voronoi diagram and Voronoi dual of a set of disks and their geometric properties. In Sect. 9.3, we describe a local-search approximation algorithm for MIN-DS of DIGs and show that it yields a PTAS. In Sect. 9.4, we present a two-stage approximation algorithm for MIN-CDS of DIGs.

9.2 Voronoi Diagram and Dual of Disks

A pair of disk disks are said to be *geometrically redundant* if one is contained in the other. A set of four disks form a *degenerate quadruple* if there is a circle which is either externally tangent to all of them (see Fig. 9.1a) or internally tangent to all of them (see Fig. 9.1b).

Let $\mathcal{D}$ be a finite set of disks in which no pair of disk are geometrically redundant and no quadruple of disk are degenerate. Then the centers of the disk in $\mathcal{D}$ are all distinct. Let V be the set of centers of the disk in $\mathcal{D}$. For $v \in V$, we use $D(v)$ to denote the disk in $\mathcal{D}$ centered at v and $\rho(v)$ to denote the radius of the disk $D(v)$. The *shifted distance* from a point p and a node $v \in V$ is defined to be

$$\ell(p,v) = \|pv\| - \rho(v)$$

For a point p and a node $v \in V$, denote

$$\ell(p,v) = \|pv\| - \rho(v)$$

In other words, $|\ell(p,v)|$ is the Euclidean distance from p to the boundary of the disk $D(v)$, and $\ell(p,v)$ is positive (respectively, negative) if p is outside (respectively, inside) $D(v)$. Figure 9.2 illustrates the shifted distances. Clearly, for each point p and any two nodes u and v in V, if $\ell(p,u) \leq \ell(p,v)$ and $p \in D(v)$, then $p \in D(u)$ as well. For each $v \in V$, the set of points p in the plane satisfying that

$$\ell(p,v) = \min_{u \in V} \ell(p,u)$$

is referred to the *Voronoi cell* of $D(v)$. The lemma below shows that the Voronoi cell of $D(v)$ is nonempty and is star-shaped with respect to v.

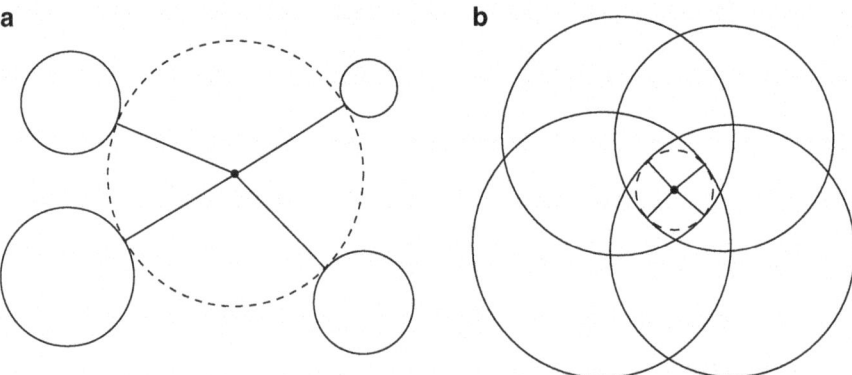

Fig. 9.1 Degenerate quadruples

9.2 Voronoi Diagram and Dual of Disks

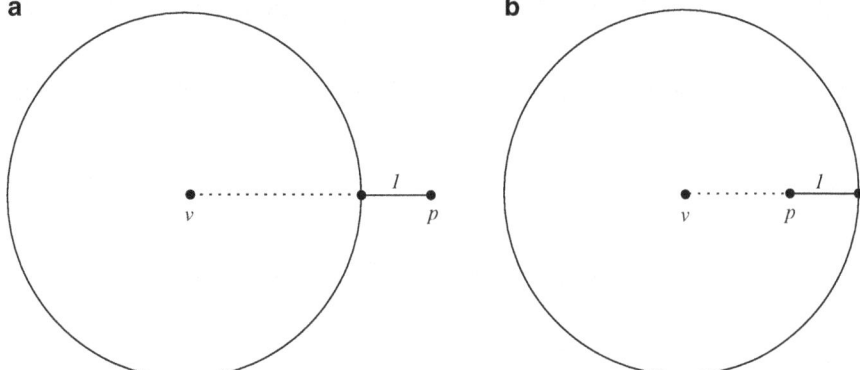

Fig. 9.2 The shifted distance

Lemma 9.2.1. *Consider any $v \in V$.*

1. *v lies in only the Voronoi cell of $D(v)$.*
2. *For any point p in the cell of v, each point in the interior of the line segment vp lies in only the Voronoi cell of $D(v)$.*

Proof. (1) For any $u \in V \setminus \{v\}$,

$$\ell(v,u) - \ell(v,v) = \|vu\| - (\rho(u) - \rho(v)) > 0,$$

where the last inequality follows from the fact that $D(u)$ and $D(v)$ are not geometrically redundant. Thus, the first part of the lemma holds.

(2) Consider any point q in the interior of the line segment vp and any $u \in V \setminus \{v\}$. We have

$$\begin{aligned}
\ell(q,v) &= \|qv\| - \rho(v) \\
&= \|pv\| - \|pq\| - \rho(v) \\
&= \ell(p,v) - \|pq\| \\
&\leq \ell(p,u) - \|pq\| \\
&= \|pu\| - \|pq\| - \rho(u) \\
&\leq \|qu\| - \rho(u) \\
&= \ell(q,u).
\end{aligned}$$

We further claim that $\ell(q,v) \neq \ell(q,u)$. Assume to the contrary that the claim does not hold. Then,

$$\|pu\| - \|pq\| = \|qu\|$$

and
$$\ell(p,v) = \ell(p,u).$$

So, q lies in the line segment pu. By symmetry, we assume that v also lies in the segment pu. Then,

$$\begin{aligned}\|uv\| &= \|pu\| - \|pv\| \\ &= (\ell(p,u)+\rho(u)) - (\ell(p,v)+\rho(v)) \\ &= \rho(u) - \rho(v).\end{aligned}$$

This means that $D(v)$ is internally tangent to $D(u)$, which is a contradiction. Thus, our claim holds. Therefore,

$$\ell(q,v) < \ell(q,u).$$

So, the second part of the lemma holds. □

Clearly, the boundary of the Voronoi cell of each disk in $\mathcal{D}$ is a concatenation of parts of hyperbolic curves and/or lines. The Voronoi cells of all disks in $\mathcal{D}$ induce a decomposition of the plane, which is known as the *Voronoi diagram* of $\mathcal{D}$. Since $\mathcal{D}$ contains no degenerate quadruple, no point belongs to Voronoi cells of more than three disks in $\mathcal{D}$. A vertex of the Voronoi diagram of $\mathcal{D}$ is an point which belongs to the Voronoi cells of three disks in V. The *Voronoi dual* of $\mathcal{D}$ is a graph on V in which two nodes u and v are adjacent if and only if the Voronoi cells of $D(u)$ and $D(v)$ share a common point. It is a planar graph as shown in the lemma below.

Lemma 9.2.2. *The Voronoi dual of $\mathcal{D}$ is a planar graph.*

Proof. Consider any edge $e = uv$ of the Voronoi dual of $\mathcal{D}$. Let p_e be an arbitrary common point shared by the Voronoi cells of $D(u)$ and $D(v)$ which is not a vertex of the Voronoi diagram of $\mathcal{D}$. The poly-segment up_ev, which is the concatenation of the two line segments up_e and vp_e, is referred to as the geometric embedding of e in the plane. We show that the geometric embeddings of any two edges e and e' do not cross each other (i.e., have no common interior point). Assume to the contrary that they have a common interior point q. We consider in two cases.

Case 1: e and e' have no common endpoint. Let $e = uv$ and $e' = u'v'$. By Lemma 9.2.1, any interior point of the poly-segment up_ev other than p_e either lies only in the Voronoi cell of $D(u)$ or only in the Voronoi cell of $D(v)$, and hence cannot lie in poly-segment $u'p_{e'}v'$. Thus, q must be the point p_e. Similarly, q must be the point $p_{e'}$. However, $q = p_e = p_{e'}$ would imply that $\{u,v,u',v'\}$ is a degenerate quadruple, which is a contradiction.

Case 2: e and e' have one common endpoint. Let $e = uv$ and $e' = u'v$. By Lemma 9.2.1, any interior point of the line segment up_e lies only in the Voronoi cell of $D(u)$, and hence cannot lie in poly-segment $u'p_{e'}v$. Thus, q must lie in the line segment vp_e. Similarly, q must lie in the line segment $vp_{e'}$. However, $p_e \neq p_{e'}$ for otherwise, p_e would be a vertex of the Voronoi diagram of $\mathcal{D}$, which contradicts

9.3 Local Search for MIN-DS   155

to the selection of p_e. Thus, the two line segments vp_e and $vp_{e'}$ only meet at v. So, $q = v$, which is a contradiction.

In either case, we have reached a contradiction. So, the geometric embeddings of any two edges e and e' do not have a cross each other. Therefore, the lemma holds.
□

9.3 Local Search for MIN-DS

In this section, we present a local-search algorithm for MIN-DS. Suppose that each node in V has a unique ID for tie-breaking. A node $v \in V$ is said to be *redundant* if there exists a node $u \in V$ satisfying that either v only dominates a proper subset of nodes dominated by u, or v dominates exactly the same set of nodes but has a larger ID than u. Let V^* denote the set of nonredundant nodes in V. Clearly, V^* still contains a minimum DS. Let B be a DS contained in V^*. A set $U \subseteq B$ is said to be a *loose* subset of B if there is a subset U' of V^* such that $|U'| < |U|$ and $(B \setminus U) \cup U'$ is still a DS, and to be a *tight* subset of B otherwise. B is said to be k-*tight* if every subset $U \subseteq B$ with $|U| \leq k$ is tight. Intuitively, for sufficiently large k the size of a k-tight DS is close to the domination number γ, which is the size of a minimum DS. Technically, we relate a k-tight DS with a minimum DS using the following planar expansion theorem established in [83].

Theorem 9.3.1. *There are two fixed positive constants c and K such that for any planar bipartite graph $H = (X, Y; E)$ satisfying that $|X| \geq 2$ and for every subset $Y' \subseteq Y$ of size at most $k \geq K$, $|N_H(Y')| \geq |Y'|$, we have*

$$|Y| \leq (1 + c/\sqrt{k})|X|.$$

With the help of the above theorem, we shall prove the following relation between the size of k-tight DS and the domination number γ.

Theorem 9.3.2. *Let c and K be the two fixed constants in Theorem 9.3.1. Then, for any k-tight DS $B \subseteq V^*$ with $k \geq \max\{K, 2\}$,*

$$|B| \leq \left(1 + c/\sqrt{k}\right)\gamma.$$

Theorem 9.3.2 suggests a local-search algorithm for MIN-DS, referred to as k-*Local Search* (k-**LS**), where k is a positive integer parameter at least two. It computes a k-tight cover $B \subseteq V'$ in two phases:

- *Preprocessing Phase*: Compute the set V^* of nonredundant nodes in V, and then compute a cover $B \subseteq V^*$ by the well-known greedy algorithm for Minimum Set Cover.

- *Replacement Phase*: While B is not k-tight, find a subset U of B with size at most k and a subset U' of V^* with size at most $|U|-1$ satisfying that $(B\setminus U)\cup U'$ is still a DS; replace B by $(B\setminus U)\cup U'$. Finally, we output B.

By Theorem 9.3.2, the algorithm k-**LS** has an approximation ratio at most $1 + O\left(1/\sqrt{k}\right)$ when $k \geq K$. Its running time is dominated by the second phase. Let $m = |V^*|$. Then, the second phase consists of $O(m)$ iterations. In each iteration, the search for the subset U and its replacement U' takes at most

$$O\left(m^k\right) \cdot O\left(m^{k-1}\right) = O\left(m^{2k-1}\right)$$

time. So, the total running time is

$$O(m) \cdot O\left(m^{2k-1}\right) = O\left(m^{2k}\right).$$

This means that the algorithm k-**LS** is a PTAS.

We move on to the proof of 9.3.2. Consider a minimum DS O contained in V^*. Theorem 9.3.2 holds trivially if $|B| = |O|$. So, we assume that $|B| > |O|$. Then,

$$|B\setminus O| > |O\setminus B|.$$

In addition, $|O\setminus B| \geq k$ for otherwise, we can choose a subset of $|O\setminus B|+1$ nodes from $B\setminus O$ and replace them by $O\setminus B$ to get a smaller DS, which contradicts to the fact that B is k-tight. Let T be the set of nodes in V not dominated by $O\cap B$. Then, each node in T is dominated by some node in $B\setminus O$ and by some node in $O\setminus B$. In addition, we have the following stronger property.

Lemma 9.3.3. *There is a planar bipartite graph H on $O\setminus B$ and $B\setminus O$ satisfying the following "locality condition": For each $t \in T$, there are two adjacent nodes in H both of which dominate t.*

Let H be the planar bipartite graph satisfying the property in the above lemma. We claim that for any $U \subseteq B\setminus O$, $(B\setminus U)\cup N_H(U)$ is still a DS. Indeed, consider any $t \in V$. If t is dominated by $B\setminus U$, then it is also dominated by $(B\setminus U)\cup N_H(U)$. If t is not dominated by $B\setminus U$, then t is only dominated by nodes in U and hence $t \in T$. By Lemma 9.3.3, there exist two adjacent nodes $u \in B\setminus O$ and $v \in O\setminus B$ both of which dominate t. Then, we must have $u \in U$ and hence $v \in N_H(U)$. Thus, t is still dominated by $(B\setminus U)\cup N_H(U)$. So, the claim holds.

Now, consider any $U \subseteq B\setminus O$ with $|U| \leq k$. Then $|N_H(U)| \geq |U|$, for otherwise $(B\setminus U)\cup N_H(U)$ is a DS smaller than B, which contradicts to the fact that B is k-tight. By Theorem 9.3.1, we have

$$|B\setminus O| \leq (1+c/\sqrt{k})|O\setminus B|$$

and hence

$$|B| \leq (1+c/\sqrt{k})|O|.$$

So, Theorem 9.3.2 holds.

9.3 Local Search for MIN-DS

In the remaining of this section, we prove Lemma 9.3.3. Let B' (respectively, O') be a replication of $B \setminus O$ (respectively, $O \setminus B$), and let $V' = O' \cup B'$. Each replication $v \in V'$ also has a radius $r(v)$ equal to the radius of the original node in V it is replicated from. For each $v \in V'$, define

$$\bar{r}(v) = \min\{\|uv\| - r(u) : \|uv\| > r(u) + r(v), u \in V\}.$$

Clearly, $\bar{r}(v) > r(v)$, and if we increase the radius of v to any value below $\bar{r}(v)$, the set of nodes in V dominated by v remains the same. A function ρ on V' is said to be *domination-preserving* if $r(v) \leq \rho(v) < \bar{r}(v)$ for each $v \in V'$. For each domination-preserving function ρ, we use $\mathcal{D}_\rho$ to denote the collection of disks centered at v of radius $\rho(v)$ for all $v \in V'$.

Lemma 9.3.4. *There exists a domination-preserving function ρ on V' such that $\mathcal{D}_\rho$ contains no degenerate quadruple.*

Proof. We prove the lemma by contradiction. Assume the lemma is not true. Let ρ be the "fewest counterexample", in other words, $\mathcal{D}_\rho$ contains the least number of degenerate quadruples. Suppose that the disk centered at $u \in V'$ is contained in at least one quadruple in $\mathcal{D}_\rho$. We show that we can change the radius of u to some value in $[r(u), \bar{r}(u))$ such that the disk of u is not involved in any degenerate quadruple. Consider any triple disks D_1, D_2, D_3 in $\mathcal{D}_\rho$ which can potentially form a degenerate quadruple with some disk centered at u. Let v_i be the center of D_i for $1 \leq i \leq 3$. For each circle which is either externally tangent to the triple or internally tangent to the triple, its center q must satisfy the equalities

$$\|qv_1\| - \rho(v_1) = \|qv_2\| - \rho(v_2) = \|qv_3\| - \rho(v_3).$$

So, q lies in a branch of a hyperbola with two foci v_1 and v_2 (which can be degenerated to the perpendicular bisector of $v_1 v_2$), and similarly, q also lies in a branch of a hyperbola with two foci v_1 and v_3 (which can be degenerated to the perpendicular bisector of $v_1 v_3$). Since these two branches may have at most 4 intersection points, q can take at most 4 positions. Thus, for a disk centered at u to form a degenerate quadruple with D_1, D_2, and D_3, its radius can be of at most 4 values, each of which is referred to as a *forbidden* radius of u. As the number of triples of disks in $\mathcal{D}_\rho$ which can potentially form a degenerate quadruple with some disk centered at u is at most $\binom{|V'|-1}{3}$, the total number of forbidden radii of u is at most $4\binom{|V'|-1}{3}$. Now consider the radius function ρ' on V' satisfying that $\rho'(u)$ takes some value in $[r(u), \bar{r}(u))$ other than the forbidden radii of u, and $\rho'(v) = \rho(v)$ for each $v \neq u$. Then, ρ' is still domination-preserving but $\mathcal{D}_{\rho'}$ contains strictly fewer degenerate quadruples. This contradicts to the choice of ρ. Therefore, the lemma holds. □

Now, we fix a domination-preserving function ρ on V' such that $\mathcal{D}_\rho$ contains no degenerate quadruple. For each node $v \in V'$, let $D(v)$ denote the disk centered at v of radius $\rho(v)$. We claim that any pair of disks in $\mathcal{D}_\rho$ are geometrically nonredundant.

Indeed, assume to the contrary that there exist two nodes in u and v such that $D(u) \subseteq D(v)$. Since ρ is domination-preserving, all nodes in V dominated by u are also dominated by v, which is a contradiction. Thus, our claim holds. Let H be the graph obtained from the Voronoi dual of $\mathcal{D}_\rho$ by removing all edges between two nodes in O' and all edges between two nodes in B'. By Lemma 9.2.2, H is a planar bipartite graph on O' and B'.

Next, we show that H satisfies the locality condition: For each $t \in T$, there are two adjacent nodes in H both of which dominate t. Clearly, t is dominated by a node $v \in V'$ if and only if $\ell(t,v) \leq \rho(t)$ where $\ell(t,v) = \|tv\| - \rho(v)$ is the shifted distance from t to v. Thus, if $\ell(t,u) \leq \ell(t,v)$ for some two nodes u and v in V' and t is dominated by v, then t is also dominated by u as well. We consider two cases:

Case 1: t lies in the Voronoi cell of $D(u)$ for some $u \in O'$. Then, u must dominate t as t is dominated by O'. Let v be a node in B' to which t has the smallest shifted distance. Then, v must also dominate t, as t is dominated by B'. If u and v are adjacent, then the locality condition holds trivially. So, we assume that u and v are nonadjacent. Then, t lies outside the Voronoi cell of $D(v)$. We walk from t to v along the straight line segment tv. During this walk, we may cross some Voronoi cells of the disks in $\mathcal{D}_\rho$, and at some point before reaching v we will enter the Voronoi cell of $D(v)$ the first time. Let x be the point at which we first enter the Voronoi cell of $D(v)$. We must enter this cell from another cell, and we assume this cell the Voronoi cell of $D(w)$. Then, $\ell(t,w) \leq \ell(t,v)$ as

$$\begin{aligned}\ell(t,w) &= \|tw\| - \rho(w) \\ &\leq \|tx\| + \|xw\| - \rho(w) \\ &= \|tx\| + \ell(x,w) \\ &= \|tx\| + \ell(x,v) \\ &= \|tx\| + \|xv\| - \rho(v) \\ &= \|tv\| - \rho(v) \\ &= \ell(t,v).\end{aligned}$$

We further claim that $\ell(t,w) < \ell(t,v)$. Indeed, assume to the contrary that $\ell(t,w) = \ell(t,v)$. Then, we must have $\|tw\| = \|tx\| + \|xw\|$, in other words, w lies in the ray tv. As $\ell(t,w) = \ell(t,v)$, either $D(v) \subseteq D(w)$ or $D(w) \subseteq D(v)$, which is a contradiction. Therefore, our claim is true. By the choice of v, $w \in O'$ and w is adjacent to v. In addition, w dominates t since $\ell(t,w) < \ell(t,v)$ and v dominates t. Thus, the locality condition is satisfied.

Case 2: t lies in the Voronoi cell of $D(u)$ for some $u \in B'$. The proof is the same as in Case 1 is thus omitted.

Since ρ is domination-preserving and B' (respectively, O') be a replication of $B \setminus O$ (respectively, $O \setminus B$), Lemma 9.3.3 holds.

9.4 A Two-Staged Algorithm for MIN-CDS

In this section, we present a two-staged approximation algorithm for MIN-CDS of DIGs. The first stage applies the local-search algorithm k-**LS** presented in the previous section to compute a DS B. The second stage compute a set C of connectors such that $B \cup C$ is a CDS as follows. Initially C is empty. Repeat the following iteration until $B \cup C$ is connected. First we find a pair of closest connected components of $G_r(B \cup C)$ and compute a shortest (in terms the number of hops) path P between them. Then, we all internal nodes in P to C.

Clearly, the number of iterations executed in the second stage is at most $|B|-1$. In addition, it is easy to show that at most two nodes are added to C in each iteration. Thus, We claim that $|C| \leq 2(|B|-1)$. So, $|B \cup C| \leq 3|B|-2$. By Theorem 9.3.2,

$$|B| = \left(1 + O\left(1/\sqrt{k}\right)\right)\gamma.$$

Therefore,

$$|B \cup C| = \left(3 + O\left(1/\sqrt{k}\right)\right)\gamma.$$

Since γ is no more than the connected domination number γ_c, the two-staged approximation algorithm has an approximation bound $3 + O\left(1/\sqrt{k}\right)$.

Chapter 10
Geometric Hitting Set and Disk Cover

> *Another means of silently lessening the inequality of property is to exempt*
> *all from taxation below a certain point, and to tax the higher portions of*
> *property in geometric progression as they rise.*
>   THOMAS JEFFERSON

10.1 Motivation and Overview

MIN-SENSOR-COVER is a special case of MIN-SET-COVER, which can be seen as MIN-SET-COVER in a geometric case with the base set formed by all targets and all given subsets of targets induced by sensing disks. When all sensing disks have the same size, a classic result indicates that MIN-SENSOR-COVER has PTAS. In this chapter, we introduce some related results in case that sensing disks may have different sizes. Those results may lead us to a sequence of research works on coverage and connected coverage with different sizes of sensing disks.

10.2 Minimum Geometric Hitting Set

Consider a set V of nodes and a set $\mathcal{D}$ of target disks in the plane. A node $v \in V$ is said to *hit* a disk $D \in \mathcal{D}$ if $v \in D$. A subset S of V is said to be a *hitting set* (HS) of $\mathcal{D}$ if each disk in $\mathcal{D}$ is hit by some node in V. The problem of finding a minimum subset of V which is an HS of $\mathcal{D}$ is referred to as MIN-HITTING-SET. In this section, we present a PTAS for MIN-HITTING-SET.

If a target disk is hit by all nodes in V, we simply remove it from $\mathcal{D}$. In addition, if a target disk contains some other target disk, we also remove it from $\mathcal{D}$. Thus, we assume that each target disk in $\mathcal{D}$ is not hit by at least one node in V and does not contain any other disk in $\mathcal{D}$. Consequently, the disks in $\mathcal{D}$ have distinct centers. Let T denote the set of the centers of disks in $\mathcal{D}$. For each $t \in T$, we use $D(t)$ to denote the disk in $\mathcal{D}$ centered at t, and $r(t)$ to denote the radius of the disk $D(t)$. Let S be an HS. A set $U \subseteq S$ is said to be a *loose* subset of S if there is a subset U' of V such that $|U'| < |U|$ and $(S \setminus U) \cup U'$ is still an HS, and to be a *tight* subset of S otherwise. S is said to be k-*tight* if every subset $U \subseteq S$ with $|U| \leq k$ is tight. Intuitively, a k-tight HS for sufficiently large k is close to the minimum HS in size. We will formerly prove such relation in the next theorem.

Theorem 10.2.1. *Let c and K be the two universal constants in Theorem 9.3.1. Then, for any k-tight hitting set $S \subseteq V'$ with $k \geq \max\{K, 2\}$, $|C| \leq \left(1 + c/\sqrt{k}\right)$ opt, where opt is the size of a minimum hitting set.*

Theorem 10.2.1 suggests a local search algorithm for MIN-DISK-COVER, referred to as k-*Local Search* (k-**LS**), where k is a positive integer parameter at least two. It computes a k-tight HS S in two phases:

- *Preprocessing Phase.* Compute an HS S by the well-known greedy algorithm for minimum set cover.
- *Replacement Phase.* While S is not k-tight, find a subset U of S with size at most k and a subset U' of V with size at most $|U| - 1$ satisfying that $(S \setminus U) \cup U'$ is still an HS; replace S by $(S \setminus U) \cup U'$. Finally, we output S.

By Theorem 10.2.1, the algorithm k-**LS** has an approximation ratio at most $1 + O\left(1/\sqrt{k}\right)$ when $k \geq K$. Its running time is dominated by the second phase. Let $m = |V|$. Then, the second phase consists of $O(m)$ iterations. In each iteration, the search for the subset U and its replacement U' takes at most

$$O\left(m^k\right) \cdot O\left(m^{k-1}\right) = O\left(m^{2k-1}\right)$$

time. So, the total running time is

$$O(m) \cdot O\left(m^{2k-1}\right) = O\left(m^{2k}\right).$$

This means that the algorithm k-**LS** is a PTAS.

We move on to the proof of 10.2.1. Let O be a minimum HS. Theorem 10.2.1 holds trivially if $|S| = |O|$. So, we assume that $|S| > |O|$. Let $S' = S \setminus O$ and $O' = O \setminus S$. Then, $|S'| > |O'|$. In addition, $|O'| \geq k$ for otherwise, we can choose a subset of $|O'| + 1$ nodes from S' and replace them by O' to get a smaller HS, which contradicts to the fact that S is k-tight. Let T' be the set of centers of the target disks not hit by $O \cap S$. Then, for each $t \in T'$, $D(t)$ is hit by some node in S' and by some node in O'. In addition, we have the following stronger property.

Lemma 10.2.2. *There is a planar bipartite graph H on O' and S' satisfying the following "locality condition": For each $t \in T$, there are two adjacent nodes in H, both of which hit $D(t)$.*

10.2 Minimum Geometric Hitting Set

Let H be the planar bipartite graph satisfying the property in the above lemma. We claim that for any $U \subseteq S'$, $(S \setminus U) \cup N_H(U)$ is still an HS. Indeed, consider any $t \in T$. If $D(t)$ is hit by $S \setminus U$, then it is also hit by $(S \setminus U) \cup N_H(U)$. If $D(t)$ is not hit by $S \setminus U$, then $D(t)$ is only hit by nodes in U and hence $t \in T'$. By Lemma 10.2.2, there exist two adjacent nodes $u \in S'$ and $v \in O'$, both of which hit $D(t)$. Then, we must have $u \in U$ and hence $v \in N_H(U)$. Thus, $D(t)$ is still hit by $(S \setminus U) \cup N_H(U)$. So, the claim holds.

Now, consider any $U \subseteq S'$ with $|U| \leq k$. Then $|N_H(U)| \geq |U|$, for otherwise $(S \setminus U) \cup N_H(U)$ is an HS smaller than S, which contradicts to the fact that S is k-tight. By Theorem 9.3.1, we have

$$|S'| \leq (1 + c/\sqrt{k})|O'|$$

and hence

$$|S| \leq (1 + c/\sqrt{k})|O|.$$

So, Theorem 10.2.1 holds.

In the remaining of this section, we prove Lemma 10.2.2. Let $V' = O' \cup S'$. Consider any $t \in T'$. Define

$$\bar{r}(t) = \min_{v \in V'} \left\{ \frac{\|tv\| + r(t)}{2} : \|tv\| > r(v) \right\}.$$

Clearly, $\bar{r}(t) > r(t)$. Let $D'(t)$ be the disk centered at t of radius $\bar{r}(t)$. Then, for each node $v \in V'$, v hits $D(t)$ if and only if v hits $D'(t)$. Let $\mathcal{D}'$ to denote the collection of disks $D'(t)$ for all $t \in T'$. Consider any $v \in V'$. Define

$$\delta_1(v) = \min\{\bar{r}(t) - \|tv\| : \|tv\| \leq r(t), t \in T'\},$$
$$\delta_2(v) = \min\{\|vu\|/3 : u \in V \setminus \{v\}\},$$
$$\delta(v) = \min\{\delta_1(v), \delta_2(v)\}.$$

Then, $\delta(v) > 0$ and the disk of radius $\delta(v)$ centered at v is referred to as the *perturbation range* of v. Then, for any point v' in the perturbation range of v and any $t \in T'$, v hits $D(t)$ if and only if v' hits $D'(t)$. In addition, the perturbation ranges of all nodes in V' are disjoint. A *restricted perturbation* of V' is a mapping σ from V' to the plane such that for each $v \in V'$, $\sigma(v)$ lies within the perturbation range of v.

A set of four points in the plane form a *degenerate quadruple* if they all lie on some circle. The next lemma shows that V' has a restricted perturbation containing no degenerate quadruple.

Lemma 10.2.3. *There exists a restricted perturbation σ of V' such that $\sigma(V')$ contains no degenerate quadruple.*

Proof. We prove the lemma by contradiction. Assume the lemma is not true. Let σ be the "fewest counterexample," in other words, $\sigma(V')$ contains the least number of degenerate quadruples. Suppose that a node $\sigma(u) \in V'$ is contained in at least one degenerate quadruple in $\sigma(V')$. We show that we can change $\sigma(u)$ to some point in the perturbation range of u which is not involved in any degenerate quadruple. For any triple nodes $\{v_1, v_2, v_3\}$ in $V' \setminus \{u\}$ such that $\sigma(v_1)$, $\sigma(v_2)$, and $\sigma(v_3)$ are not collinear, the circumcircle of $\{\sigma(v_1), \sigma(v_2), \sigma(v_3)\}$ is referred to as a *forbidden circle* of u. As the number of forbidden circles of u is at most $\binom{|V'|-1}{3}$, there is a point u' which lies in the perturbation range of u but not on any forbidden circle of u. Let σ' be the restricted perturbation of $O' \cup S'$ obtained from σ by replacing $\sigma(u)$ with u'. Then, $u' = \sigma'(u)$ is not contained in any degenerate quadruple of $\sigma'(V')$. Thus, $\sigma'(V')$ contains strictly fewer degenerate quadruples than $\sigma(V')$, which contradicts to the choice of σ. Therefore, the lemma holds. □

Now, we fix a restricted perturbation σ of V' satisfying that $\sigma(V')$ contains no degenerate quadruple. Let G' be the graph obtained from Voronoi dual of $\sigma(V')$ by removing all edges between two nodes in O' and all edges between two nodes in S'. Then, G' is planar. In the next, we show that G' satisfies the locality condition: For each target $t \in T'$, there are two adjacent nodes in G', both of which hit $D'(t)$. We consider two cases:

Case 1: t lies in the Voronoi cell of $\sigma(u)$ for some $u \in O'$. Then, $\sigma(u)$ must hit $D'(t)$ as $D'(t)$ is hit by $\sigma(O')$. Let v be a node in S' such that $\sigma(v)$ has the shortest distance from t. Then, $\sigma(v)$ must also hit $D'(t)$ as $D'(t)$ is hit by $\sigma(O')$. If $\sigma(u)$ and $\sigma(v)$ are adjacent, then lemma holds trivially. So, we assume that $\sigma(u)$ and $\sigma(v)$ are nonadjacent. Then t lies outside the Voronoi cell of $\sigma(v)$. We walk from t to $\sigma(v)$ along the straight line segment $t\sigma(v)$. During this walk, we may cross some Voronoi cells, and at some point before reaching $\sigma(v)$, we will enter the Voronoi cell of $\sigma(v)$ the first time. Let x be the point at which we first enter the Voronoi cell of $\sigma(v)$. We must enter this cell from another cell, and we assume the cell is the Voronoi cell of $\sigma(w)$. Then, $\sigma(w)$ does not lie in the ray $x\sigma(v)$, and hence

$$\|t\sigma(w)\| < \|tx\| + \|x\sigma(w)\| = \|tx\| + \|x\sigma(v)\| = \|t\sigma(v)\|.$$

Since $\sigma(v)$ hits $D'(t)$, $\sigma(w)$ hits $D'(t)$ as well; and by the choice of v, $w \in O'$. As $\sigma(w)$ is adjacent to $\sigma(v)$, the locality condition is satisfied.

Case 2: t lies in the Voronoi cell of $\sigma(u)$ for some $u \in S'$. The proof is the same as in Case 1 and is thus omitted.

Finally, we define a graph H on V' such that two nodes u and v are adjacent if and only if $\sigma(u)$ and $\sigma(v)$ are adjacent in G'. Then, H is also a planar bipartite graph. In addition, for any target $t \in T'$, let $\sigma(u)$ and $\sigma(v)$ be two adjacent nodes in G', both of which hit $D'(t)$. Then, u and v are two adjacent nodes in H, both of which hit $D(t)$. This completes the proof of Lemma 10.2.2.

10.3 Minimum Disk Cover

Consider a set $\mathcal{D}$ of disks and a set T of target points in the plane. A disk $D \in \mathcal{D}$ is said to *cover* a target $t \in T$ if $t \in D$. A subset $\mathcal{D}'$ of $\mathcal{D}$ is said to be a *cover* of T if each target in T is covered by some node in V. The problem of finding a minimum subset of $\mathcal{D}$ which is a cover of T is referred to as MIN-DISK-COVER. In this section, we present a PTAS for MIN-DISK-COVER.

Suppose that each disk in $\mathcal{D}$ has a unique ID for tie-breaking. A disk $D \in \mathcal{D}$ is said to be *redundant* if there exists another disk $D' \in \mathcal{D}$ satisfying that either D only covers a proper subset of targets covered by D', or D covers exactly the same set of targets as D' but has a larger ID than D. If a disk in $\mathcal{D}$ is redundant, we simply remove it from $\mathcal{D}$. Thus, we assume that no disk in $\mathcal{D}$ is redundant. Consequently, we can identify the disks in $\mathcal{D}$ by their centers. Let V denote the set of the centers of disks in $\mathcal{D}$. For each $v \in V$, we use $D(v)$ to denote the disk in $\mathcal{D}$ centered at v, and $r(v)$ to denote the radius of the disk $D(v)$. For simplicity, a node $v \in V$ is said to *cover* a target $t \in T$ if $D(v)$ covers t, a subset C of V is said to be a *cover* of T if the set of disks $\{D(v) : v \in C\}$ is a cover of T.

Let $C \subseteq V$ be a cover of T. A set $U \subseteq C$ is said to be a *loose* subset of C if there is a subset U' of V such that $|U'| < |U|$ and $(C \setminus U) \cup U'$ is still a cover, and to be a *tight* subset of C otherwise. C is said to be *k-tight* if every subset $U \subseteq C$ with $|U| \leq k$ is tight. Intuitively, a k-tight cover for sufficiently large k is close to the minimum cover in size. We will formerly prove such relation in the next theorem.

Theorem 10.3.1. *Let c and K be the two universal constants in Theorem 9.3.1. Then, for any k-tight cover $C \subseteq V'$ with $k \geq \max\{K, 2\}$, $|C| \leq \left(1 + c/\sqrt{k}\right) opt$, where opt is the size of a minimum cover.*

Theorem 10.3.1 suggests a local search algorithm for MIN-DISK-COVER, referred to as *k-Local Search* (*k*-**LS**), where k is a positive integer parameter at least two. It computes a k-tight cover C in two phases:

- *Preprocessing Phase.* Compute a cover $C \subseteq V$ by the well-known greedy algorithm for minimum set cover.
- *Replacement Phase.* While C is not k-tight, find a subset U of C with size at most k and a subset U of V with size at most $|U| - 1$ satisfying that $(C \setminus U) \cup U'$ is still a cover; replace C by $(C \setminus U) \cup U'$. Finally, we output C.

By Theorem 10.3.1, the algorithm *k*-**LS** has an approximation ratio at most $1 + O\left(1/\sqrt{k}\right)$ when $k \geq K$. Its running time is dominated by the second phase. Let $m = |V|$. Then, the second phase consists of $O(m)$ iterations. In each iteration, the search for the subset U and its replacement U' takes at most

$$O\left(m^k\right) \cdot O\left(m^{k-1}\right) = O\left(m^{2k-1}\right)$$

time. So, the total running time is

$$O(m) \cdot O\left(m^{2k-1}\right) = O\left(m^{2k}\right).$$

This means that the algorithm k-**LS** is a PTAS.

We move on to the proof of 10.3.1. Let $O \subseteq V$ be a minimum cover. Theorem 10.3.1 holds trivially if $|C| = |O|$. So, we assume that $|C| > |O|$. Let $C' = C \setminus O$ and $O' = O \setminus C$. Then, $|C'| > |O'|$. In addition, $|O'| \geq k$ for otherwise, we can choose a subset of $|O'| + 1$ nodes from C' and replace them by O' to get a smaller cover, which contradicts to the fact that C is k-tight. Let T' be the set of targets not covered by $O \cap C$. Then, each $t \in T'$ is covered by some node in O' and by some node in C'. In addition, we have the following stronger property.

Lemma 10.3.2. *There is a planar bipartite graph H on O' and C' satisfying the following "locality condition": For each $t \in T$, there are two adjacent nodes in H, both of which cover t.*

Let H be the planar bipartite graph satisfying the property in the above lemma. We claim that for any $U \subseteq C'$, $(C \setminus U) \cup N_H(U)$ is still a cover. Indeed, consider any $t \in T$. If t is covered by $C \setminus U$, then it is also covered by $(C \setminus U) \cup N_H(U)$. If t is not covered by $C \setminus U$, then it is only covered by nodes in U and hence $t \in T'$. By Lemma 10.3.2, there exist two adjacent nodes $u \in C'$ and $v \in O'$, both of which cover t. Then, we must have $u \in U$ and hence $v \in N_H(U)$. Thus, t is still covered by $(C \setminus U) \cup N_H(U)$. So, the claim holds.

Now, consider any $U \subseteq C'$ with $|U| \leq k$. Then $|N_H(U)| \geq |U|$, for otherwise $(C \setminus U) \cup N_H(U)$ is a cover smaller than C, which contradicts to the fact that C is k-tight. By Theorem 9.3.1, we have

$$|C'| \leq (1 + c/\sqrt{k})|O'|$$

and hence

$$|C| \leq (1 + c/\sqrt{k})|O|.$$

So, Theorem 10.3.1 holds.

In the remaining of this section, we prove Lemma 10.3.2. Let $V' = O' \cup C'$. For each $v \in V'$, define

$$\bar{r}(v) = \min_{t \in T} \{\|tv\| : \|tv\| > r(v)\},$$

Clearly, $\bar{r}(v) > r(v)$, and if we increase the radius of v to any value below $\bar{r}(v)$, the set of targets covered by v remains the same. A function ρ on V' is said to be *coverage-preserving* if $r(v) \leq \rho(v) < \bar{r}(v)$ for each $v \in V'$. For each coverage-preserving function ρ, we use $\mathcal{D}_\rho$ to denote the collection of disks centered at v of radius $\rho(v)$ for all $v \in V'$.

Lemma 10.3.3. *There exists a coverage-preserving function ρ on V' such that $\mathcal{D}_\rho$ contains no degenerate quadruple.*

10.3 Minimum Disk Cover

Now, fix a coverage-preserving function ρ on V' such that $\mathcal{D}_\rho$ contains no degenerate quadruple. For each node $v \in V'$, let $D'(v)$ denote the disk centered at v of radius $\rho(v)$. We claim that any pair of disks in $\mathcal{D}_\rho$ are geometrically nonredundant. Indeed, assume to the contrary that there exist two nodes in u and v such that $D'(u) \subseteq D'(v)$. Since ρ is coverage-preserving, all targets covered by u are also covered by v, which is a contradiction. Thus, our claim holds. Let H be the graph obtained from the Voronoi dual of $\mathcal{D}_\rho$ by removing all edges between two nodes in O' and all edges between two nodes in C'. Then, H is a planar bipartite graph on O' and C'.

Next, we show that H satisfies the locality condition: For each $t \in T$, there are two adjacent nodes in H, both of which cover t. Clearly, t is covered by a node $v \in V'$ if and only if $\ell(t,v) \leq 0$ where $\ell(t,v) = \|tv\| - \rho(v)$ is the shifted distance from t to v. Thus, if $\ell(t,u) \leq \ell(t,v)$ for some two nodes u and v in V' and t is covered by v, then t is also covered by u. We consider two cases:

Case 1: t lies in the Voronoi cell of $D'(u)$ for some $u \in O'$. Then, u must cover t as t is covered by O'. Let v be a node in C' to which t has the smallest shifted distance. Then, v must also cover t, as t is covered by C'. If u and v are adjacent, then the locality condition holds trivially. So, we assume that u and v are nonadjacent. Then, t lies outside the Voronoi cell of $D'(v)$. We walk from t to v along the straight line segment tv. During this walk, we may cross some Voronoi cells of the disks in $\mathcal{D}_\rho$, and at some point before reaching v we will enter the Voronoi cell of $D'(v)$ the first time. Let x be the point at which we first enter the Voronoi cell of $D'(v)$. We must enter this cell from another cell, and we assume this cell the Voronoi cell of $D'(w)$. Then, $\ell(t,w) \leq \ell(t,v)$ as

$$\ell(t,w) = \|tw\| - \rho(w)$$
$$\leq \|tx\| + \|xw\| - \rho(w)$$
$$= \|tx\| + \ell(x,w)$$
$$= \|tx\| + \ell(x,v)$$
$$= \|tx\| + \|xv\| - \rho(v)$$
$$= \|tv\| - \rho(v)$$
$$= \ell(t,v).$$

We further claim that $\ell(t,w) < \ell(t,v)$. Indeed, assume to the contrary that $\ell(t,w) = \ell(t,v)$. Then, we must have $\|tw\| = \|tx\| + \|xw\|$, in other words, w lies in the ray tv. As $\ell(t,w) = \ell(t,v)$, either $D'(v) \subseteq D'(w)$ or $D'(w) \subseteq D'(v)$, which is a contradiction. Therefore, our claim is true. By the choice of v, $w \in O'$ and w is adjacent to v. In addition, w covers t since $\ell(t,w) < \ell(t,v)$ and v dominates t. Thus, the locality condition is satisfied.

Case 2: t lies in the Voronoi cell of $D'(u)$ for some $u \in C'$. The proof is the same as in Case 1 and is thus omitted.

Since ρ is coverage-preserving, Lemma 10.3.2 holds.

Chapter 11
Minimum-Latency Scheduling

> *The key is not to prioritize what's on your schedule, but to schedule your priorities.*
> STEPHEN R. COVEY

11.1 Motivation and Overview

Consider a multihop wireless network in which all network nodes V lie in plane and have a unit communication radius. Its communication topology G is the unit disk graph (UDG) of V. Under the protocol interference model, every node has a communication radius normalized to one, and an interference radius ρ for some parameter $\rho \geq 1$ (see Fig. 11.1). A node v can receive the message successfully from a transmitting node u if v is within the transmission range of u but is outside the interference range of any other node transmitting simultaneously.

In this chapter, we study minimum-latency schedulings for the following four group communications in the multihop wireless network:

- *Broadcast*. A distinguished source node sends a common packet to all other nodes.
- *Data Aggregation*. A distinguished sink node collects the data aggregated from all the packets at the nodes other than the sink node. In other words, every intermediate node combines all received packet with its own packet into a single packet of fixed size according to some aggregation function such as logical and/or, maximum, or minimum.
- *Data Gathering*. A distinguished sink node collects a packet from every other node.
- *Gossiping*. Every node broadcasts a common packet to all other nodes.

Suppose that all communications proceed in synchronous time slots and a node can transmit at most one packet of a fixed size in each time slot. A communication schedule for each of these four communication tasks not only specifies a

Fig. 11.1 Each node has a unit communication radius and an interference radius $\rho \geq 1$

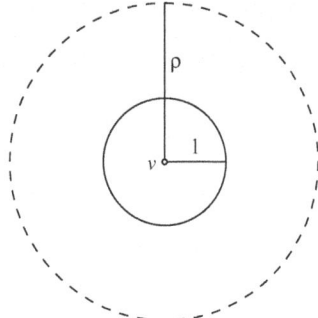

communication routing but also assigns a time slot to every communication link in the routing subject to two constraints:

1. The link ordering given by the routing should be followed.
2. All communication links assigned in each time slot are interference-free.

The latency of a communication schedule is the number of time slots during which at least one transmission occurs. The minimum-latency schedulings for the above four group communications are all NP-hard. In this chapter, we present constant-approximation algorithms for them.

The CDS plays a critical role in the design of scheduling algorithms [109, 110]. Indeed, all relaying nodes of a message must form a CDS of G. However, in order to achieve a short latency, the CDS has to be short and sparse instead of being just small. Specifically, consider a node $s \in V$ and let R be the graph radius of G with respect to s. A CDS U of G with $s \in U$ should be "short" in the sense that the graph radius of $G[U]$ with respect to s is bounded by a constant factor of R and "sparse" in the sense that the maximum degree of $G[U]$ is bounded by a small constant. In Sect. 11.3, we will present a construction of such short and sparse CDS.

The following terms and notations will be adopted throughout this chapter. Let $n = |V|$. The connected domination number of G is denoted by γ_c. The unit disk centered at a node v is denoted by $D(v)$. The topological boundary of a planar set A is denoted by ∂A. The directed version of G, denoted by $\overrightarrow{G}$, is the digraph obtained from G by replacing every edge e in G with two oppositely oriented links between the two endpoints of e. A subset U of nodes is said to be distance-d independent for some $d > 0$ if and only if their mutual Euclidean distances are greater than d. Equivalently, a set of nodes are distance-d independent if and only if they form an independent set of the d-disk graph on V. For any $d > 0$, a distance-d coloring of a subset U of nodes is an assignment of colors to the nodes in U such that any pair of nodes of distance at most d receives distinct colors. Let X and Y be two disjoint subsets of V. Y is a *cover* of X if each node in X is adjacent to some node in Y, and a *minimal cover* of X if Y is a cover of X but no proper subset of Y is a cover of X. Any ordering $y_1, y_2, \cdots, y_m$ of Y induces a minimal cover $Z \subseteq Y$ of X by the following sequential pruning method: Initially, $Z = Y$. For each $i = m$ down to 1,

11.2 Geometric Preliminaries

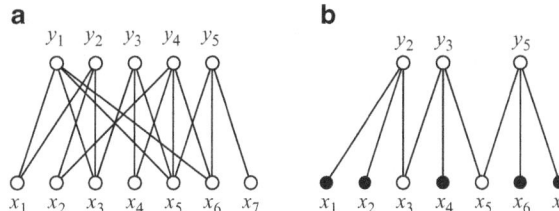

Fig. 11.2 (a) $X = \{x_i : 1 \leq i \leq 7\}$ is covered by $Y = \{y_i : 1 \leq i \leq 5\}$. (b) $\{y_2, y_3, y_4\}$ is a minimal cover of X. The *black* nodes are private neighbors

Table 11.1 Summary on the approximation bounds of the scheduling algorithms

Communication	Approximation bound
Broadcast	$2\beta_\rho$
Aggregation with $\rho > 1$	$(\alpha_\rho + 12)\beta_\rho$
Gathering	$2\beta_\rho$
Gossiping	$4\beta_\rho$

if $Z \setminus \{y_i\}$ is a cover of X, remove y_i from Z. Figure 11.2 is an illustration of such sequential pruning method. Suppose that Y is a cover of X. A node $x \in X$ is called a *private* neighbor of a node $y \in Y$ with respect to Y if y is the only node in Y which is adjacent to x. Clearly, if Y is a minimal cover of X, then each node in Y has at least one private neighbor with respect to Y.

Short and sparse CDS has important applications in the design of scheduling algorithms for group communications in wireless networks. In this chapter, we have constructed a short and sparse CDS and built a dominating tree on such CDS which is used as the routing for the group communications. By exploiting the rich structural properties of the dominating tree, we are able to design scheduling algorithms with constant approximation bounds for the group communications. Table 11.1 summarizes the approximation bounds of our scheduling algorithms described in this chapter for the four group communications.

11.2 Geometric Preliminaries

For any $\rho > 1$, let α_ρ denote the maximum number of points in a unit disk whose mutual distances are greater than $\rho - 1$. For any $\rho \geq 1$, let β_ρ denote the maximum number of points in a half disk of radius $\rho + 1$ whose mutual distances are greater than one. The upper bounds on α_ρ and β_ρ can be derived by the following classic result on disk packing.

Theorem 11.2.1 (Zassebhaus-Groemer-Oler Inequality). *Suppose that S is a compact convex set and U is a set of points with mutual distances at least one. Then*

$$|U \cap S| \leq \frac{\text{area}(S)}{\sqrt{3}/2} + \frac{\text{peri}(S)}{2} + 1,$$

where $\text{area}(S)$ *and* $\text{peri}(S)$ *are the area and perimeter of S, respectively.*

Fig. 11.3 The two circles have unit radius, and $1 \le \|uv\| \le 2$. Then, both $\triangle pvx$ and $\triangle qvy$ are equilateral

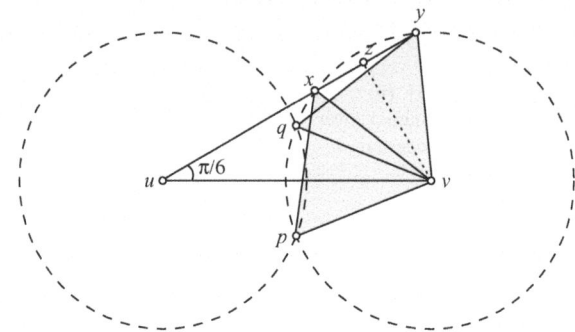

When the set S is a disk or a half disk, we have the following packing bound.

Corollary 11.2.2. *Suppose that S (respectively, S') is a disk (respectively, half disk) of radius r, and U is a set of points with mutual distances at least one. Then*

$$|U \cap S| \le \frac{2\pi}{\sqrt{3}} r^2 + \pi r + 1,$$

$$|U \cap S'| \le \frac{\pi}{\sqrt{3}} r^2 + \left(\frac{\pi}{2} + 1\right) r + 1.$$

By the above corollary,

$$\alpha_\rho \le \left\lfloor \frac{2\pi/\sqrt{3}}{(\rho-1)^2} + \frac{\pi}{\rho-1} \right\rfloor + 1$$

and

$$\beta_\rho \le \left\lfloor \frac{\pi}{\sqrt{3}}(\rho+1)^2 + \left(\frac{\pi}{2}+1\right)(\rho+1) \right\rfloor + 1.$$

Now, we introduce an interesting "equilateral triangle property," which will be used in the proof of Lemma 11.2.4.

Lemma 11.2.3. *Consider two nodes u and v with $1 \le \|uv\| \le 2$. Let p and q be their two intersection points of $\partial D(u)$ and $\partial D(v)$ (see Fig. 11.3). Suppose that x and y are the two intersection points between $\partial D(v)$ and the ray emanating from u which is apart from uv by $30°$ and is on the same side of uv as q, with x being between u and y. Then $\triangle pvx$ and $\triangle qvy$ are equilateral.*

Proof. Let z be the midpoint of xy. Then, vz is perpendicular to xy and

$$\widehat{xvz} = \arccos \|vz\| = \arccos \frac{\|uv\|}{2} = \widehat{pvu}.$$

11.2 Geometric Preliminaries

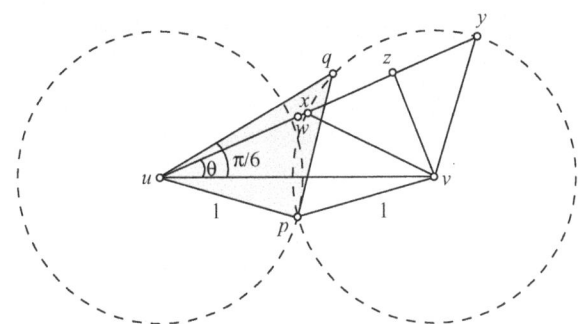

Fig. 11.4 If $\theta \leq 2\arcsin\frac{1}{4}$, then $\|uy\| \geq \|uv\|$, and hence $w \in ux \subset \triangle upq$

Hence,
$$\widehat{pvx} = \widehat{pvu} + \widehat{uvx} = \widehat{xvz} + \widehat{uvx} = \widehat{uvz} = \frac{\pi}{3}.$$

Similarly, $\widehat{yvz} = \widehat{uvq}$ and hence
$$\widehat{qvy} = \widehat{yvz} + \widehat{qvz} = \widehat{uvq} + \widehat{qvz} = \widehat{uvz} = \frac{\pi}{3}.$$

Thus, the lemma follows. □

The next lemma presents two sufficient conditions for the intersection of two unit disks being covered by a third unit-disk.

Lemma 11.2.4. *Consider three nodes u, v, and w satisfying that $1 < \|uv\| \leq 2$ and $\|vw\| > 1$. Then,*
$$D(u) \cap D(v) \subseteq D(w)$$
if one of the following two conditions holds

1. $\|uw\| \leq 1$ *and* $\widehat{vuw} \leq \frac{\pi}{6}$.
2. $1 < \|uw\| \leq \|uv\|$ *and* $\widehat{vuw} \leq 2\arcsin\frac{1}{4} \approx 28.955°$ *(see Fig. 11.4).*

Proof. Let p be the intersection point of $\partial D(u)$ and $\partial D(v)$ which lies on the different side of uv from w, and q be the point on $\partial D(v)$ satisfying that q is on the same side of uv, $\widehat{quv} = \frac{\pi}{6}$ and $\widehat{uqv} \geq \frac{\pi}{2}$ (see Fig. 11.4). By 11.2.3, $\|pq\| = 1$. We will show that w lies in $\triangle upq$. This would imply that $\|pw\| \leq 1$ and consequently
$$D(u) \cap D(v) \subseteq D(w).$$

Under the first condition in the lemma, w lies in $\triangle upq$ obviously. So we assume the second condition in the lemma holds. Let x and y be the intersection points of the ray uw and $\partial D(v)$ with x being closer to u than y (see Fig. 11.4). We claim that $\|uy\| \geq \|uv\|$. Note that
$$\widehat{uyv} = \widehat{vxy} = \widehat{xuv} + \widehat{xvu},$$
$$\widehat{uvy} = \widehat{xvu} + \widehat{xvy}.$$

It is sufficient to show that $\widehat{xvy} \geq \widehat{xuv}$. For the simplicity of presentation, we denote $\widehat{xuv}$ by θ. Let z be the midpoint of xy. Then, z is the perpendicular foot of v on uw, and

$$\begin{aligned}\|vz\| &= \|uv\| \sin\theta \leq 2\sin\theta \\ &= 4\sin\frac{\theta}{2}\cos\frac{\theta}{2} \\ &\leq 4\sin\left(\arcsin\frac{1}{4}\right)\cos\frac{\theta}{2} \\ &= \cos\frac{\theta}{2}.\end{aligned}$$

Hence,

$$\widehat{xvy} = 2\arccos\|vz\| \geq 2\arccos\left(\cos\frac{\theta}{2}\right) = \theta.$$

Thus, our claim holds. So,

$$\|uw\| \leq \|uv\| \leq \|uy\|,$$

which means w is on the line segment uy. As $w \notin D(v)$, w must be on the line segment ux. Consequently, w lies in $\triangle upq$ as well. □

11.3 Dominating Tree

In this section, we describe a rooted spanning tree T of G constructed from a connected dominating set (CDS). This tree will be used in the routings of all the four group communications. Depending on the type of the group communications, the root of T, denoted by s, is chosen as follows. For broadcast, s is the source of the broadcast; for aggregation or gathering, s is the sink node; for gossiping, s is a graph center of G. In either case, we use L to denote the graph radius of G with respect to s.

We begin with the construction of a small, short, and sparse CDS of G. We first select a maximal independent set (MIS) I of G in the first-fit manner in a breadth-first-search (BFS) ordering (with respect to s) of V. All nodes in I form a dominating set, and hence are referred to as dominators. Then, we select a set C of connectors to interconnect I as follows. Let G' be the graph on I in which there is edge between two dominators if and only if they have a common neighbor. The radius of G' with respect to s is denoted by L'. Clearly, $L' \leq L - 1$. For each $0 \leq l \leq L'$, let I_l be the set of dominators of depth l in G'. Then, $I_0 = \{s\}$. For each $0 \leq l < L'$, let P_l be the set of nodes adjacent to at least one node in I_l and at least one node in I_{l+1}, and compute a minimal cover $C_l \subseteq P_l$ of I_{l+1} (see an illustration in Fig. 11.5). Set $C = \bigcup_{l=0}^{L'-1} C_l$. Then, $I \cup C$ is a CDS of G. We will prove the following lemma on the sparsity of $I \cup C$.

11.3 Dominating Tree

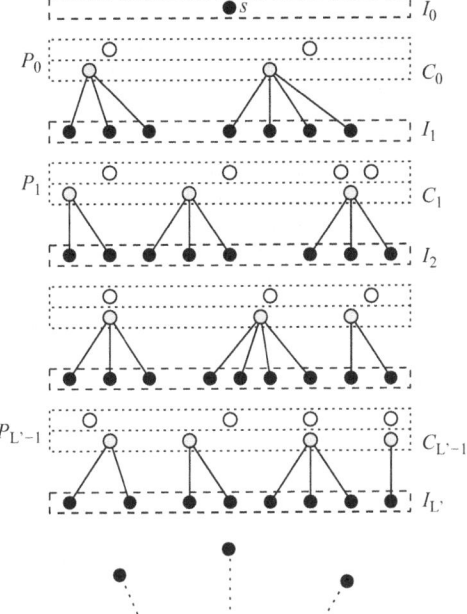

Fig. 11.5 The selection of connectors (marked by *gray*)

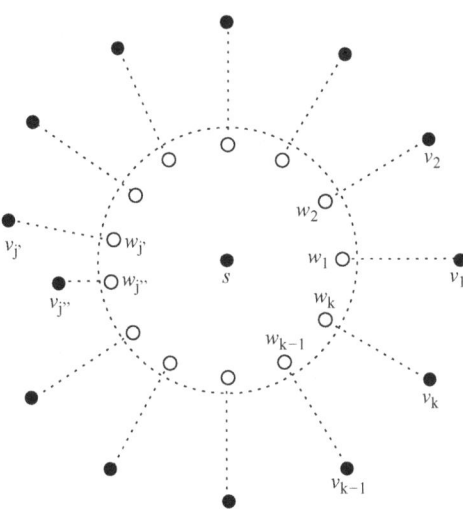

Fig. 11.6 $w_1, w_2, \ldots, w_k$ are the connectors in C_0. Each v_j is a private dominator neighbor of w_j in I_1 with respect to C_0

Lemma 11.3.1. $|C_0| \leq 12$ and each dominator in I_l with $1 \leq l \leq L' - 1$, is adjacent to at most 11 connectors in C_l.

Proof. We first prove that $|C_0| \leq 12$. Assume to the contrary that $C_0 = \{w_1, w_2, \ldots, w_k\}$ for some $k \geq 13$. By the minimality of C_0, for each $1 \leq j \leq k$, there is a node $v_j \in I_1$ such that v_j is adjacent to w_j but not to any other node in C_0 (see Fig. 11.6). Among the k nodes $v_1, \ldots, v_k$, there exist two, say $v_{j'}$ and $v_{j''}$, satisfying that $\angle v_{j'} s v_{j''} \leq \frac{2\pi}{13}$. Assume by symmetry that $v_{j''}$ is closer to s than $v_{j'}$. Since the distance between $v_{j'}$ and $v_{j''}$ is greater than one, $D(s) \cap D(v_{j'}) \subseteq D(v_{j''})$ by Lemma 11.2.4. Hence, $w_{j'} \in D(v_{j''})$, which is a contradiction.

Fig. 11.7 $w_1, w_2, \ldots, w_k$ are the connectors in C_l adjacent to u. Each v_j with $1 \leq j \leq k$ is a private dominator neighbor of w_j in I_{l+1} with respect to C_l

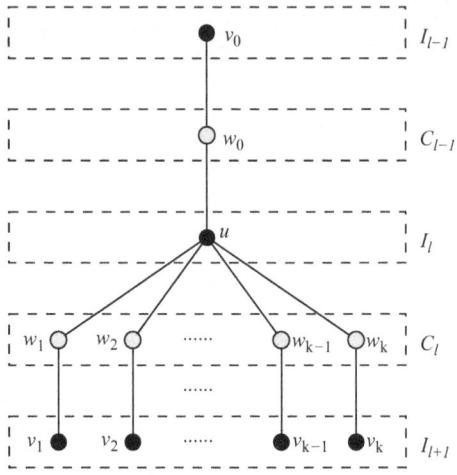

Next, we show that each dominator u in I_l for some $1 \leq l \leq L' - 1$ is adjacent to at most 11 connectors in C_l. Suppose that $w_1, w_2, \ldots, w_k$ are the connectors in C_l which are adjacent to u. By the minimality of C_l, for each $1 \leq j \leq k$, there is a node $v_j \in I_{l+1}$ such that v_j is a private neighbor of w_j with respect to C_l (see Fig. 11.7). Let w_0 be a connector in C_{l-1} which is adjacent to u, and v_0 be a dominator in I_{l-1} which is adjacent to w_0. Then, for each $0 \leq j \leq k$, w_j is the only node in $\{w_0, w_1, \ldots, w_k\}$ which is adjacent to v_j. By the same argument above, we can show that $k + 1 \leq 12$, which implies $k \leq 11$. □

Now, we construct T by specifying the parent of each node other than s. First, each dominator in I_l with $1 \leq l \leq L'$ chooses the neighboring connector of the smallest ID in C_{l-1} as its parent. Second, each connector in C_l with $0 \leq l \leq L' - 1$ chooses the neighboring dominator of the smallest ID in I_l as its parent. Third, each other node, referred to as dominate, chooses the neighboring dominator of the smallest ID as its parent. Clearly, T is a spanning tree and is called a *dominating tree*. Figure 11.8 is an illustration of the construction of T. By the property of the CDS $I \cup U$, the maximum depth of T is at most $2L' + 1 \leq 2L - 1$, s has at most 12 connector children, and each other dominator has at most 11 connector children.

In the remaining of this section, we present a first-fit distance-$(\rho + 1)$ coloring of an arbitrary subset U of dominators. In the lexicographic order of U, all nodes in U are sorted from the left to the right with ties broken by the ordering from the bottom to the top. Suppose that $\langle u_1, u_2, \ldots, u_k \rangle$ is the lexicographic order of U. The first-fit coloring in this order uses colors represented by natural numbers and runs as follows: Assign the color 1 to u_1. For $i = 2$ up to k, assign to u_i with the smallest color not used by any v_j with $j < i$ and $\|v_i v_j\| \leq (\rho + 1)$. We claim that at most β_ρ colors are used by this coloring. Indeed, consider an arbitrary node $u \in U$. All other nodes in U which precede u and are apart from u by a distance at most $\rho + 1$ lie in the left half disk of radius $\rho + 1$ centered at u. The number of these dominators is at most $\beta_\rho - 1$, where the -1 term is due to that u is also in this half disk. Hence, the color number received by u is at most $(\beta_\rho - 1) + 1 = \beta_\rho$. Thus, our claim holds.

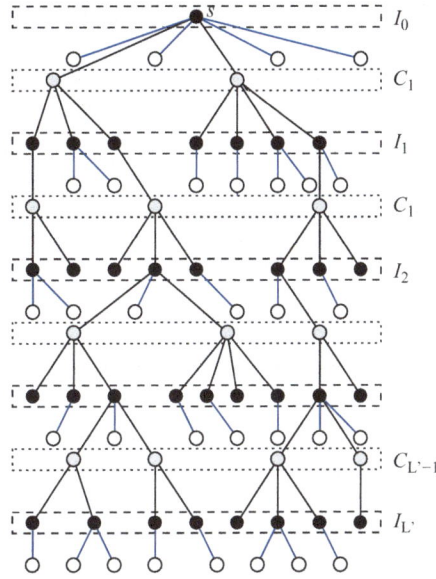

Fig. 11.8 An illustration of dominating tree

11.4 Broadcast Scheduling

Let s be the source of the broadcast. We first construct the dominating tree T rooted at s as in Sect. 11.3. The routing of the broadcast is the spanning s-aborescence oriented from T. The broadcast schedule is then partitioned in $2L' + 1$ rounds sequentially dedicated to the transmissions by

$$I_0, C_0, I_1, C_1, \ldots, I_{L'-1}, C_{L'-1}, I_{L'}$$

respectively. For each $1 \leq l \leq L'$, we compute a first-fit distance-$(\rho + 1)$ coloring of I_l in the lexicographic order. The individual rounds are then scheduled as follows:

- In the round for I_0, only the source node s transmits, and hence this round has only one time slot.
- In the round for C_0, all nodes in C_0 transmit one by one, and thus this round takes at most 12 time slots.
- In the round for I_l with $1 \leq l \leq L'$, a dominator of the ith color transmits in the ith time slot, and hence this round takes at most β_ρ time slots.
- In the round for C_l with $1 \leq l \leq L' - 1$, a connector with a child dominator of the ith color transmits in the ith time slot, and hence this round also takes at most β_ρ time-slots.

Thus, the latency of the entire broadcast schedule is at most

$$1 + 12 + (2L' - 1)\beta_\rho$$
$$\leq 13 + \beta_\rho (2L - 3)$$
$$= 2\beta_\rho L - (3\beta_\rho - 13).$$

Since L is a trivial lower bound on the minimum broadcast latency, the above broadcast schedule is a $2\beta_\rho$-approximation of the optimum.

11.5 Aggregation Scheduling

Let s be the sink of the aggregation. Let Δ denote the maximum degree of G, and L be the graph radius of G with respect to s. For the trivial case that $L = 1$, we simply let all nodes other than s transmit one by one. Such trivial schedule has latency $n - 1 = \Delta$. Subsequently, we assume that $L > 1$. We first construct the dominating tree T rooted as s as in Sect. 11.3. The routing of the aggregation schedule is the spanning inward s-aborescence oriented from T. Let W denote the set of dominates. The aggregation schedule is then partitioned in $2L' + 1$ rounds sequentially dedicated to the transmissions by

$$W, I_{L'}, C_{L'-1}, I_{L'-1}, \ldots, C_1, I_1, C_0$$

respectively. We describe a procedure used by the scheduling in the round for W and the round for each C_l with $1 \leq l \leq L' - 1$.

Let B be a set of links whose receiving endpoints are all dominators. Suppose that ϕ is the maximum number of links with a common dominator endpoint. We first partition into at most ϕ subsets B_j with $1 \leq j \leq \phi$ such that each dominator is incident to at most one link in each B_j. The schedule of B is then further partitioned into ϕ sub-rounds dedicated to $B_1, B_2, \ldots, B_\phi$, respectively. In the sub-round for B_j, we compute a first-fit distance-$(\rho + 1)$ coloring of the dominators incident to the links in B_j, and then all links in B_j whose dominator endpoints receive the ith color are scheduled in the ith time slot. Thus, each of the ϕ consists of at most β_ρ time slots. Hence, the total number of slots is at most $\phi \beta_\rho$.

Now, we are ready to describe the schedule in the individual rounds.

- In the round for W, we adopt the above procedure to produce a schedule in this round. Since each dominator is adjacent to at least one dominate, the maximum number of nodes in W adjacent to a dominator is at most $\Delta - 1$. Hence, this round takes at most $(\Delta - 1)\beta_\rho$ time slots.
- In the round for C_l with $1 \leq l \leq L' - 1$, we also adopt the above procedure to produce a schedule in this round. Since each dominator in I_{l-1} is adjacent to at most 11 connectors in C_l, this round takes at most $11\beta_\rho$ time slots.
- In the round for C_0, all nodes in C_0 transmit one by one, and thus this round takes at most 12 time slots.

- In the round for I_l with $1 \leq l \leq L'$, we compute a first-fit distance-$(\rho+1)$ coloring of I_l in the lexicographic order and let each dominator with the ith color transmit in the ith time slot. This round takes at most β_ρ time slots.

Thus, the latency of the entire aggregation schedule is at most

$$(\Delta-1)\beta_\rho + 11\beta_\rho(L'-1) + 12 + L'\beta_\rho$$
$$= \Delta\beta_\rho + 12\beta_\rho(L'-1) + 12$$
$$\leq \Delta\beta_\rho + 12\beta_\rho(L-2) + 12$$
$$= \Delta\beta_\rho + 12\beta_\rho L - 12(2\beta_\rho - 1).$$

Since the trivial case takes Δ time slots and $\beta_\rho > 1$, we have the following theorem.

Theorem 11.5.1. *The latency of the above aggregation schedule is at most $\Delta\beta_\rho + 12\beta_\rho L - 12(2\beta_\rho - 1)$.*

In the next, we present a lower bound on the minimum aggregation latency in terms of Δ.

Lemma 11.5.2. *For any $\rho > 1$, the minimum aggregation latency is at least Δ/α_ρ.*

Proof. Let u be a node with maximum degree in G, and S be the unit disk centered at u. Then, S contains $\Delta+1$ nodes. If s is not in S, then all these $\Delta+1$ nodes in S have to transmit; otherwise, exactly Δ nodes in S have to transmit. In either case, at least Δ nodes in S have to transmit. Since all nodes transmitting in the same time slot must be apart from each other by a distance greater than $\rho - 1$, at most α_ρ nodes in C can transmit in a time slot. Hence, the Δ transmissions by the nodes in S take at least Δ/α_ρ time slots. □

Since L is also a trivial lower bound on the minimum aggregation latency, the approximation bound of the aggregation schedule is at most

$$\alpha_\rho\beta_\rho + 12\beta_\rho = (\alpha_\rho + 12)\beta_\rho.$$

11.6 Gathering Scheduling

Let s be the sink of the gathering. If $L=1$, then all other nodes transmit to s one by one, and this schedule is optimal. So, we assume subsequently that $L > 1$. We first construct the dominating tree of G rooted at s. The routing of the gathering schedule is the spanning inward s-aborescence oriented from T. Our gather schedule utilizes a labelling of the edges of T, which is described below.

Let $\langle v_1, v_2, \ldots, v_{n-1} \rangle$ be an ordering of $V \setminus \{s\}$ in the descending order of depth in T with ties broken arbitrarily. For $1 \leq i \leq n$, we assign the jth edge in the tree path from s to v_j with a label $2(i-1)+j$ (see an example in Fig. 11.9). Clearly,

Fig. 11.9 A multi-labelling of the edges in the dominating tree

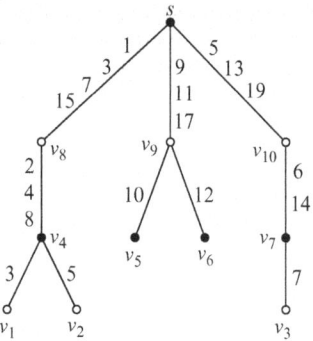

the number of labels received by an edge connecting v and its parent is equal to the number of descendents (including v itself) of v in T. If v is a connector (respectively, dominator), all labels received by the edge between v and its parent are odd (respectively, even). In addition, all edges across two consecutive layers of the dominating tree receive distinct labels. We further claim that the largest label is $2n-3$. Consider a node v_i and let h be the length of the path from s to v_i. The maximum label assigned to the edges in the path from s to v_i is $2(i-1)+h$. It is sufficient to show that

$$2(i-1)+h \leq 2n-3.$$

Since none of $v_1, v_2, \ldots, v_{i-1}$ belongs to the path from s to v_i, we have

$$h+i-1 \leq n-1.$$

and hence $i \leq n-h$. Therefore,

$$2(i-1)+h \leq 2(n-h-1)+h = 2n-h-2 \leq 2n-3.$$

So, the claim holds.

For each $1 \leq k \leq 2n-3$, let E_k denote the set of edges of T which has been assigned with a label k, and A_k denote the links in the inward s-arborescence oriented from the edges in E_k. Then, for odd (respectively, even) k, all the receiving (respectively, transmitting) endpoints of links in A_k are dominators. In addition, for each $1 \leq k \leq 2n-3$, every dominator is incident to at most one link in A_k.

Now, we are ready to describe the gathering schedule. The schedule is partitioned in $2n-3$ rounds sequentially dedicated to

$$A_{2n-3}, A_{2n-2}, \ldots, A_2, A_1$$

respectively. For each $1 \leq k \leq 2n-3$, the round for A_k is scheduled as follows. We first compute a first-fit distance-$(\rho+1)$ coloring of the dominator endpoints of the links A_k. Then each link whose dominator endpoint receives the ith color is

| dominator subframe | connector subframe | dominator subframe | connector subframe | dominator subframe | connector subframe |

Fig. 11.10 Framing of the time slots

scheduled in the ith time slot of the kth round. Thus, each round takes at most β_ρ time slots. Consequently, the latency of the gathering schedule is $\beta_\rho (2n-3)$. So, we have the following theorem.

Theorem 11.6.1. *The latency of the above gathering schedule is at most* $\beta_\rho (2n-3)$.

Since $n-1$ is a trivial lower bound on the minimum gathering latency, the approximation ratio of the gathering schedule presented in this section is at most $2\beta_\rho$.

11.7 Gossiping Scheduling

Let s be the graph center of G. If $L=1$, we adopt the following two-phased schedule. In the first phase, all nodes other than s transmit one by one. This phase takes $n-1$ time slots. In the second phase, the source node transmit all the received packets and its own packet one by one. This phase takes n time slots. So, the total latency is $2n-1$. Clearly, n is a trivial lower bound on the minimum gossiping latency, as every node has to transmits at least once and receive at least $n-1$ times. Thus, its approximation factor is at most 2.

From now on, we assume that $L>1$. Our gossiping schedule consists of two phases. In the first phase s collects all the packets from all other nodes, and in the second phase s broadcasts all the n packets to all other nodes. We adopt the gathering schedule presented in the previous section for the first phase. In the sequel, the node s disseminates all received packets and its own packet to all other nodes. We present a schedule for the second phase in the next.

We first construct the dominating tree T of G rooted at s. The routing of the second phase is the spanning s-aborescence oriented from T. Then, we compute the first-fit coloring distance-$(\rho+1)$ coloring of dominators. Let k be the number of colors used by this coloring. Then, $k \leq \beta_\rho$. By proper renumbering of the colors, we assume that s has the first color. We group the time slots into $2k$-slot frames (see Fig. 11.10). In each frame, the first k slots form a dominator subframe, and the remaining k slots form a connector subframe. Only dominators (respectively, connectors) are allowed to transmit in the dominator (respectively, connector) subframe in each frame. Each dominator with color i is only allowed to transmit in the ith slot of a dominator (respectively, connector) subframe. Each connector is only allowed to transmit in the subsets of time slots corresponding to the colors of its child dominators. The source node s transmits one packet in each frame.

Each connector receiving a packet in a dominator subframe transmits the received packet in all the time slots corresponding to the colors of its child dominators of the connector subframe of the same frame. Each dominator with color i receiving a packet in a connector subframe transmits the received packet in the ith time slot of the dominator subframe of the subsequent frame.

The correctness of the above schedule is obvious. Next, we bound the latency of the second phase. After $n-1$ frames, s transmits the last packet. After another L' frames, the last packet reaches all nodes in $I_{L'}$. Finally, after another half frame, the last packet reaches all nodes. So, the total number of time slots taken by the second phase is at most

$$2k(n-1+L')+k$$
$$\leq 2k(n+L-2)+k$$
$$= 2k(n+L-1.5)$$
$$\leq 2\beta_\rho(n+L-1.5).$$

By Theorem 11.6.1, the first phase takes at most $\beta_\rho(2n-3)$ time slots. Hence, the total number of time slots taken by the two phases is at most

$$\beta_\rho(2n-3)+2\beta_\rho(n+L-1.5)$$
$$= \beta_\rho(4n-6+2L).$$

Therefore, we have the following theorem.

Theorem 11.7.1. *The latency of the two-phased gossiping schedule is at most* $\beta_\rho(4n-6+2L)$.

In the next, we present a lower bound on the minimum gossiping latency.

Lemma 11.7.2. *The minimum gossiping latency of G is at least $n-1+L$.*

Proof. The broadcasting of each message requires at least L transmissions. So, the total number of transmissions in any gossiping schedule is at least nL. This implies that some node must take at least L transmissions. On the other hand, every node must take $n-1$ receptions. Therefore, some node takes at least $n-1+L$ transmissions and receptions. This implies that $n-1+L$ is a lower bound on the minimum gossiping latency. □

Since
$$4n-6+2L = 4(n-1+L)-2(L+1) < 4(n-1+L).$$

Therefore, the approximation factor of our gossiping schedule is at most $4\beta_\rho$.

Chapter 12
CDS in Planar Graphs

> *Simple, geometric forms and planar surfaces*
> *define Jeep Patriot's timeless, purpose-built design.*
> TREVOR CREED

12.1 Motivation and Overview

Although MIN-CDS in general graphs is hard to approximate, the restriction to certain special graph classes admits much better approximation results. MIN-CDS in planar graphs remains NP-hard even for planar graphs that are regular of degree 4 [57]. The related problem, MIN-DS in planar graphs, is also NP-hard even for planar graphs with maximum vertex degree 3 and planar graphs that are regular of degree 4 [57]. It is well known that MIN-DS in planar graphs possesses a polynomial-time approximation scheme (PTAS) based on the shifting strategy [3]: For any constant $\varepsilon > 0$, there is a polynomial-time $(1 + \varepsilon)$-approximation algorithm. Thus, it is immediate to conclude that MIN-CDS in planar graphs can be approximated within a factor $3 + \varepsilon$ for any $\varepsilon > 0$ in polynomial time. However, the degree of the polynomial grows with $1/\varepsilon$ and hence, the approximation scheme is hardly practical.

In this chapter, we present a simple heuristic for MIN-CDS in general graphs developed in [105]. When running on graphs excluding K_m (the complete graph of order m) as a minor, the heuristic has an approximation ratio of at most 7 if $m = 3$, or at most $\frac{m(m-1)}{2} + 5$ if $m \geq 4$. In particular, if running on a planar graph, the heuristic has an approximation ratio of at most 15. The remaining of this chapter is organized as follows. In Sect. 12.2, we introduce some related graph-theoretic concepts and parameters. In Sect. 12.3, we describe the heuristic for MIN-CDS in general graphs. In Sect. 12.4, we provide an upper bound on the cardinality of the CDS output by the heuristic.

12.2 Preliminaries

Let $G = (V, E)$ be a graph. We sometimes write $V(G)$ instead of V and $E(G)$ instead of E. For any $U \subseteq V$, we use $G[U]$ to denote the subgraph of G induced by U. The distance $\text{dist}_G(u, v)$ in G of two vertices $u, v \in V(G)$ is the length of a shortest path between u and v in G. The distance between a vertex v and a set $U \subseteq V(G)$ is

$$\min_{u \in U} \text{dist}_G(u, v).$$

The distance between two subsets U and W of $V(G)$ is

$$\min_{u \in U} \min_{w \in W} \text{dist}_G(u, w).$$

A vertex set $U \subseteq V(G)$ is a k-independent set (k-IS) of G if the distance between any pair of vertices in U is greater than k. The k-independence number of G, denoted by $\alpha_k(G)$, is the largest cardinality of a k-IS. Note that a 1-IS is a usual IS and $\alpha_1(G)$ is the usual independence number $\alpha(G)$. The domination number of G, denoted by $\gamma(G)$, the connected domination number of G, denoted by $\gamma_c(G)$, and $\alpha_2(G)$ are related by the following inequality [44].

$$\alpha_2(G) \leq \gamma(G) \leq \gamma_c(G).$$

To see why $\alpha_2(G) \leq \gamma(G)$, let $U \subseteq V(G)$ be a maximum 2-IS of G. For each $u \in U$, let $N_G[u]$ denote the closed neighborhood of u in G. Then the closed neighborhoods $N_G[u]$ for all $u \in U$ are pairwise disjoint. Thus, each dominating set of G must contain at least one vertex from each $N_G[u]$. This implies that $\gamma(G) \geq \alpha_2(G)$.

A *contraction* of an edge (u, v) in G is made by identifying u and v with a new vertex whose neighborhood is the union of the neighborhoods of u and v (with resulting multiple edges and self-loops deleted). A *contraction* of G is a graph obtained from G by a sequence of edge contractions. A graph H is a *minor* of G if H is the contraction of a subgraph of G. G is H-*free* if G has no minor isomorphic to H. For example, by Kuratowski's theorem, a graph is planar if and only if it is both K_5-free and $K_{3,3}$-free. In this chapter, we focus on K_m-free graphs. Our algorithm would find a CDS of size at most

$$\left(\frac{m(m-1)}{2} + 5\right)\alpha_2(G) - 5$$

of a K_m-free graph G for any $m \geq 4$. This implies that if G is K_m-free for some $m \geq 4$, then

$$\gamma_c(G) \leq \left(\frac{m(m-1)}{2} + 5\right)\alpha_2(G) - 5.$$

In particular, for a planar graph G,

$$\gamma_c(G) \leq 15\alpha_2(G) - 5.$$

12.3 Algorithm Description

We first give a brief overview on the algorithm design. The algorithm is presented as a color-marking and time-stamping process. Each vertex maintains one of the three colors: black, gray, and white, which is initially white. In addition, each vertex maintains a set of stamps, which is initially empty. The algorithm runs in proceeds in iterative phases. In the kth phase, a subset B_k of nonblack vertices are marked with black, and all of their gray neighbors are stamped with the phase number (interpreted as the time) k while keeping all previous stamps, and all of their white neighbors are marked gray and stamped with the current phase number k. At the end of the kth phase, all black nodes have to be connected, and each white vertex, if there is any left, has a gray neighbor with time-stamp j for every $1 \leq j \leq k$. The algorithm ends when no white vertex is left and outputs all black vertices which form a CDS.

Now, we describe the algorithm. For the simplicity of description, we introduce some new terms and notations. Given a color marking of all vertices of G, the *deficiency graph* is the graph obtained from G by first removing all black vertices and those gray vertices without white neighbors, and then removing edges between gray vertices. Thus, each vertex of a deficiency graph is either white or gray, and each connected component of a deficiency graph must have at least one white vertex. Given a vertex v and a positive integer k, we use **MarkStamp**(v,k) to denote the basic operation which marks v black and all white neighbors of v gray and stamps v and all its nonblack neighbors with k.

Consider a connected graph H and a positive integer k which satisfy the following properties: Each vertex of H is either white or gray and at least one vertex is white. If $k = 1$, then all vertices are white; and otherwise, every white vertex is adjacent to a gray vertex stamped with j for every $1 \leq j \leq k-1$. Such pair (H,k) is referred to as a *residue pair*. A *restricted connected 2-dominating set* (RC2DS) of a residue pair (H,k) is a subset of vertices U of H satisfying that:

- $H[U]$ is connected.
- Every white vertex not in U, if there is any, is at a distance of exactly two from U.
- And for every $1 \leq j \leq k-1$, at least one vertex in U has a stamp j.

We present a simple procedure, called **RC2DS**(H,k), which takes a residue pair (H,k) as input and produces a RC2DS for (H,k) which are marked black and a color marking and time-stamping of the remaining vertices. The procedure **RC2DS**(H,k) consists of four steps:

- Step 1: Initialization. If $k \geq 2$, let $a_j = 0$ for $j = 1,\ldots,k-1$.
- Step 2: Sorting. Build a spanning tree T of H rooted at a white vertex, and compute a breadth-first-search order $v_1, v_2, \ldots, v_s$ of all white vertices in H with respect to T.
- Step 3: Coloring and Stamping. **MarkStamp**(v_1,k). For $i = 2$ to s, if v_i is white and has no gray neighbors stamped with k, proceed as follows:

(a) Set $l = 1$, $u_1 = v_i$. Repeat the following iteration until u_l is black: If $k \geq 2$ and u_l is gray, set $a_j = 1$ for each stamp $j < k$ of u_l. If u_l has a black neighbor, set u_{l+1} to any such neighbor; otherwise, if u_l has a gray neighbor stamped with k, set u_{l+1} to any such neighbor; otherwise, set u_{l+1} to its parent in T. Increment l by 1.
(b) Repeat the following iteration until $l = 1$: Decrement l by 1 and invoke **MarkStamp**(u_l, k).

- Step 4: Post-processing. If $k \geq 2$, perform the following processing. For $j = 1$ to $k - 1$, if $a_j = 0$, choose a gray neighbor u of v_1 stamped with j, set $a_t = 1$ for each stamp $t < k$ of u, and then **MarkStamp**(u, k).

The $k - 1$ boolean variables a_j for $1 \leq j \leq k - 1$ indicate whether at least one black vertex has a stamp j. They are initialized to zero in Step 1. Whenever a gray vertex with stamp j is marked black at Step 3 or Step 4, a_j is set to one. Step 4 ensures that all these boolean variables are one eventually.

The for-loop in Step 3 guarantees that in the end, every white vertex is adjacent to a gray vertex stamped with k, and thus is exactly two hops away from some black vertex. The inner loop in Step 3(a) establishes a path from a white vertex v_i without gray neighbors stamped with k to some black vertex. Let P_i be the subpath of this path excluding the black end-vertex. The inner loop in Step 3(b) invokes **MarkStamp**(u, k) for all vertices in P_i. We claim that P_i consists of either three or four vertices. Indeed, u_1 is v_i, and since v_i is white and has no gray neighbors stamped with k, u_2 is always set to the parent of v_i. Depending on the color of u_2, we consider two cases:

Case 1: u_2 is white. Then u_2 must have a gray neighbor stamped with k as early as when u_2 is examined, for otherwise, it would have been marked black. Thus, u_3 is a gray neighbor of u_2 stamped with k, and hence P_i consists of the three vertices u_1, u_2 and u_3.

Case 2: u_2 is gray. Then every stamp of u_2 is less than k. We further consider two subcases.

Subcase 2.1: At least one gray neighbor of u_2 has stamp k. Then u_3 is one of such gray neighbors and P_i just consists of the three vertices u_1, u_2 and u_3.

Subcase 2.2: None of the gray neighbors of u_2 has stamp k. As u_2 is not adjacent to any black vertex, u_3 is the parent of u_2. Since no gray vertices with stamps less than k are adjacent in H, u_3 must be white. Then at least a gray neighbor of u_3 is stamped with k as early as when u_3 is examined, for otherwise, u_3 would have been marked black. Thus, u_4 is one of such gray neighbors, and P_i just consists of the four vertices u_1, u_2, u_3, and u_4.

In summary, the path consists of either three vertices or four vertices. Furthermore, if the path consists of four vertices, then k must be greater than one and at least one a_j is set to one for some $1 \leq j \leq k - 1$ in Step 3(a).

Now we are ready to describe the algorithm, denoted by **MarkStamp**(G), for finding a CDS of G. Initially, $k = 0$, and all vertices of G have white colors. Repeat the following iteration while there are some white vertices left:

- Increment k by 1 and construct the deficiency graph G_k.
- For each connected component H of G_k, apply **RC2DS**(H,k).

Let B denote the set of black vertices produced by **MarkStamp**(G). It is easy to see that B is a CDS of G. In the next section, we will provide an upper bound on $|B|$ if the graph G is free of K_m-minor for some $m \geq 3$.

12.4 Performance Analysis

The main theorem of this section is given below.

Theorem 12.4.1. *Suppose that G is free of K_m-minor for some $m \geq 3$. If $m = 3$, then*

$$|B| \leq 7\alpha_2(G) - 4.$$

If $m \geq 4$, then

$$|B| \leq \left(\frac{m(m-1)}{2} + 5\right)\alpha_2(G) - 5.$$

By Kuratowski's theorem, a planar graph has no K_5-minor. So we have the following corollary of Theorem 12.4.1.

Corollary 12.4.2. *If G is a planar graph, then*

$$|B| \leq 15\alpha_2(G) - 5.$$

Since

$$\alpha_2(G) \leq \gamma(G) \leq \gamma_c(G),$$

Theorem 12.4.1 implies that when running on a graph G excluding K_m as a minor, the algorithm **MarkStamp**(G) has an approximation ratio of at most 7 if $m = 3$ or at most $\frac{m(m-1)}{2} + 5$ if $m \geq 4$. In particular, if running on a planar graph, the algorithm has an approximation ratio of at most 15. The remaining of this section is dedicated to the proof for Theorem 12.4.1.

Let H be a graph in which every vertex is either white or gray and there is at least one white vertex. A *restricted 2-independent set* (R2IS) is a 2-IS of H which consists of only white vertices. The *restricted 2-independence number* of H, denoted by $\alpha'_2(H)$, is the largest cardinality of an R2IS of H. Obviously, $\alpha'_2(H) \leq \alpha_2(H)$. The next lemma presents the "monotonic" properties of the deficiency graphs.

Lemma 12.4.3. *Suppose that **MarkStamp**(G) runs in l iterations. Then*

$$G = G_1 \supset G_2 \supset \cdots \supset G_l;$$
$$\alpha_2(G) = \alpha'_2(G_1) \geq \alpha'_2(G_2) \geq \cdots \geq \alpha'_2(G_l).$$

Proof. It is obvious that $G_1 = G$ and $\alpha_2'(G_1) = \alpha_2(G)$. Fix a k between 1 and $l-1$. We prove that $G_{k+1} \subset G_k$ and $\alpha_2'(G_{k+1}) \leq \alpha_2'(G_k)$.

We first show that $V(G_{k+1}) \subset V(G_k)$. Note that all white vertices of G_{k+1} must have been white in the previous iteration and thus are white vertices of G_k as well. In addition, all gray vertices of G_{k+1} which are white in the previous iteration must be white vertices of G_k. So it is sufficient to show that each gray vertex of G_{k+1} which is also gray in the previous iteration is also a vertex of G_k. Let v be a gray vertex of G_{k+1} which is also gray in the previous iteration. Then v has a white neighbor, denoted by u, in G_{k+1}. Since u is also a white vertex of G_k, v must be also a gray vertex of G_k.

Next, we show that $E(G_{k+1}) \subset E(G_k)$. Consider any edge uv of G_{i+1}. Then at least one of its endpoints is white. By symmetry, assume v is white. If u is also white, then the edge uv also appears in G_k. If u is gray, then u is either white or gray in G_k. In either case, the edge uv appears in G_k.

Finally, we show that $\alpha_2'(G_{k+1}) \leq \alpha_2'(G_k)$. Let w_1 and w_2 be any pair of white nodes of G_{k+1}. As G_{k+1} is a subgraph of G_k,

$$\text{dist}_{G_{k+1}}(w_1, w_2) \geq \text{dist}_{G_k}(w_1, w_2).$$

We claim that, however, if $\text{dist}_{G_k}(w_1, w_2) \leq 2$, then

$$\text{dist}_{G_{k+1}}(w_1, w_2) = \text{dist}_{G_k}(w_1, w_2).$$

The claim is true if $\text{dist}_{G_k}(w_1, w_2) = 1$. So we assume that $\text{dist}_{G_k}(w_1, w_2) = 2$. Then $\text{dist}_{G_{k+1}}(w_1, w_2) \geq 2$. Let v be a common neighbor of w_1 and w_2 in G_k. Then v must remain as a vertex of G_{k+1}, for otherwise, v would have been marked black in the previous iteration and both w_1 and w_2 would have become gray in G_{k+1}. Thus, $\text{dist}_{G_{k+1}}(w_1, w_2) = 2$. So our claim is true. From the claim, we conclude that if $\text{dist}_{G_{k+1}}(w_1, w_2) > 2$, then $\text{dist}_{G_k}(w_1, w_2) > 2$. This implies that $\alpha_2'(G_{k+1}) \leq \alpha_2'(G_k)$. □

The lemma below gives an upper bound on the total number of iterations if the graph G is free of K_m-minor.

Lemma 12.4.4. *If G is free of K_m-minor for some $m \geq 3$, then* **MarkStamp**(G) *runs in at most $m-1$ iterations.*

Proof. We prove the lemma by contradiction. Assume that G is free of K_m-minor but **MarkStamp**(G) runs in at least m iterations. Let H_m^* be an arbitrary connected component of G_m. By Lemma 12.4.3, for each $1 \leq k \leq m-1$, G_k has a unique connected component, denoted by H_k^*, which contains H_m^* as a subgraph. Obviously,

$$H_1^* \supset H_2^* \supset \cdots \supset H_m^*.$$

For each $1 \leq k \leq m$, let B_k^* be the set of black vertices of H_k^* marked by the procedure **RC2DS**(H_k^*, k). Then for any $1 \leq i < j \leq m$, B_i^* and B_j^* are disjoint and separated

12.4 Performance Analysis

by one hop as at least one vertex in B_j^* has a stamp i. Since each B_k^* is connected, the m sets $B_1^*, B_2^*, \ldots, B_m^*$ give rise to a K_m-minor in G, which is a contradiction. Thus, the lemma holds. □

The next lemma provides an upper bound on the number of black vertices produced by the procedure **RC2DS**(H,k).

Lemma 12.4.5. *The number of black vertices produced by the procedure* **RC2DS**(H,k) *is at most* $3\alpha_2'(H) - 2$ *if* $k = 1$, *and at most* $4\alpha_2'(H) + k - 4$ *if* $k \geq 2$.

Proof. Let $v_1, v_2, \ldots, v_s$ be the ordering of the white vertices of H produced by Step 2 of the procedure **RC2DS**(H,k). Let I be the set of integers i in $\{2, \ldots, s\}$ such that when v_i is examined in the for-loop of Step 3, v_i is white and has no gray neighbors stamped with k. It is obvious that $\{v_i : i \in \{1\} \cup I\}$ form an R2IS of H. Thus,

$$1 + |I| \leq \alpha_2'(H).$$

Next, we count the number of vertices marked black during each iteration i with $i \in I$ in the for-loop of Step 3. Fix an $i \in I$. From the explanation after the procedure **RC2DS**(H,k) in the previous section, either three or four vertices are marked black during iteration i. In addition, if four vertices are marked black in this iteration, then k must be greater than one and at least one a_j is set to one for some $1 \leq j \leq k-1$.

Finally, we count the total number of black vertices. Note that v_1 is always marked black. If for each $i \in I$, the iteration i of the for-loop at Step 3 marks exactly three vertices black, then Step 4 marks at most $k - 1$ additional vertices black. So the total number of black vertices is at most

$$1 + 3|I| + k - 1$$
$$= 3(1 + |I|) + k - 3$$
$$\leq 3\alpha_2'(H) + k - 3.$$

If for some $i \in I$, the iteration i of the for-loop at Step 3 marks four vertices black, then $k > 1$ and Step 4 marks at most $k - 2$ additional vertices black. So the total number of black vertices is at most

$$1 + 4|I| + k - 2$$
$$= 4(1 + |I|) + k - 5$$
$$\leq 4\alpha_2'(H) + k - 5.$$

Thus, if $k = 1$, the total number of black vertices is at most

$$3\alpha_2'(H) + 1 - 3 = 3\alpha_2'(H) - 2.$$

If $k \geq 2$, the total number of black vertices is at most

$$\max\left\{3\alpha_2'(H)+k-3, 4\alpha_2'(H)+k-5\right\}$$
$$\leq 4\alpha_2'(H)+k-4.$$

Therefore, the lemma holds. □

The next lemma gives upper bounds on the number of black vertices produced in each iteration of **MarkStamp**(G).

Lemma 12.4.6. *Let B_k be the set of black vertices produced in the kth iteration of **MarkStamp**(G). Then*

$$|B_1| \leq 3\alpha_2(G) - 2,$$
$$|B_2| \leq 4\alpha_2(G) - 2,$$
$$|B_3| \leq 4\alpha_2(G) - 1,$$
$$|B_k| \leq k\alpha_2(G), \quad k \geq 4.$$

Proof. From Lemmas 12.4.5 and 12.4.3, $|B_1| \leq 3\alpha_2(G) - 2$. So we assume that $k > 1$. Suppose that G_k has t connected components, denoted by $H_{k,1}, \ldots, H_{k,t}$. Since each connected component contains at least one white vertex,

$$1 \leq t \leq \sum_{i=1}^{t} \alpha_2'(H_{k,i}) = \alpha_2'(G_k).$$

For each $1 \leq i \leq t$, let $B_{k,i}$ be the vertices of $H_{k,i}$ produced by the procedure **RC2DS**$(H_{k,i}, k)$. Then

$$B_k = B_{k,1} \cup \cdots \cup B_{k,t};$$

and by Lemma 12.4.5,

$$|B_{k,i}| \leq 4\alpha_2'(H_{k,i}) + k - 4$$

for each $1 \leq i \leq t$. Thus, if $k = 2$ or 3, by Lemma 12.4.3, we have

$$|B_k| = \sum_{i=1}^{t} |B_{k,i}|$$
$$\leq 4\sum_{i=1}^{t} \alpha_2'(H_{k,i}) + (k-4)t$$
$$= 4\alpha_2'(G_k) + (k-4)t$$
$$\leq 4\alpha_2(G) + (k-4).$$

12.4 Performance Analysis

If $k \geq 4$, by Lemma 12.4.3 we have

$$\begin{aligned}
|B_k| &= \sum_{i=1}^{t} |B_{k,i}| \\
&\leq 4 \sum_{i=1}^{t} \alpha_2'(H_{k,i}) + (k-4)t \\
&\leq 4\alpha_2'(G_k) + (k-4)\alpha_2'(G_k) \\
&= k\alpha_2'(G_k) \\
&\leq k\alpha_2(G).
\end{aligned}$$

So, the lemma holds. □

Now we are ready to give the proof of Theorem 12.4.1. By Lemma 12.4.4, the total number of iterations is at most $m-1$. If $m = 3$, then by Lemma 12.4.6,

$$|B| \leq (3\alpha_2(G) - 2) + (4\alpha_2(G) - 2) \leq 7\alpha_2(G) - 4.$$

If $m = 4$, then by Lemma 12.4.6,

$$\begin{aligned}
|B| &\leq (7\alpha_2(G) - 4) + (4\alpha_2(G) - 1) \\
&= 11\alpha_2(G) - 5 \\
&= \left(\frac{m(m-1)}{2} + 5\right)\alpha_2(G) - 5.
\end{aligned}$$

If $m > 4$, by Lemma 12.4.6,

$$\begin{aligned}
|B| &\leq (11\alpha_2(G) - 5) + \sum_{k=4}^{m-1} k\alpha_2(G) \\
&= 11\alpha_2(G) - 5 + \left(\frac{m(m-1)}{2} - 6\right)\alpha_2(G) \\
&= \left(\frac{m(m-1)}{2} + 5\right)\alpha_2(G) - 5.
\end{aligned}$$

This completes the proof of Theorem 12.4.1. □

References

1. Akyildiz, I.F., Pompili, D., Melodia, T.: Underwater acoustic sensor networks: research challenges. Ad Hoc Network **3**(3), 257–279 (2005)
2. Ambühl, C., Erlebach, T., Mihalák, M., Nunkesser, M.: Constant-approximation for minimum-weight (connected) dominating sets in unit disk graphs. Proceedings of the 9th International Workshop on Approximation Algorithms for Combinatorial Optimization (APPROX 2006). Lecture Notes in Computer Science, vol. 4110 pp. 3–14. Springer, Berlin (2006)
3. Baker, B.S.: Approximation algorithms for NP-complete problems on planar graphs. J. ACM **41**(1), 153–180 (1994)
4. Baudis, G., Gröpl, C., Hougardy, S., Nierhoff, T., Prömel, H.J.: Approximating minimum spanning sets in hypergraphs and polymatroids. Technical Report, Humboldt-Universität zu Berlin (2000)
5. Benini, L., Castelli, G., Macii, A., Poncino, M., Scarsi, R.: A discrete-time battery model for high-level power estimation. Proceedings of DATE, pp. 35–39 (2000)
6. Berman, P., Calinescu, G., Shah, C., Zelikovsky, A.: Power efficient monitoring management in sensor networks. IEEE Wireless Communication and Networking Conference (WCNC'04), Atlanta, pp. 2329–2334 (2004)
7. Berman, P., Calinescu, G., Shah, C., Zelikovsky, A.: Efficient energy management in sensor networks. In: Xiao, Y., Pan, Y. (eds.) Ad Hoc and Sensor Networks, Wireless Networks and Mobile Computing, vol. 2. Nova Science Publishers, New York (2005)
8. Bharghavan, V., Das, B.: Routing in ad hoc networks using minimum connected dominating sets. International Conference on Communication, Montreal, Canada (1997)
9. Butenko, S., Kahruman-Anderoglu, S., Ursulenko, O.: On connected domination in unit ball graphs. Optim. Lett. **5**(2), 195–205 (2011)
10. Byrka, J., Grandoni, F., Rothvoss, T., Sanita, L.: An improved LP-based approximation for Steiner tree. STOC'10, pp. 583–592 (2010)
11. Calinescu, G., Ellis, R.: On the lifetime of randomly deployed sensor networks. DialM-POMC the Fifth ACM SIGACTSIGOPS International Workshop on Foundation of Mobile Computing (2008)
12. Calinescu, G., Kapoor, S., Olshevsky, A., Zelikovsky, A.: Network lifetime and power assignment in ad-hoc wireless networks. Proceedings of European Symposium on Algorithms (ESA'03), Lecture Notes in Computer Science, vol. 2832, pp. 114–126 (2003)
13. Cardei, M., Du, D.-Z.: Improving wireless sensor network lifetime through power aware organization. ACM Wireless Network. **11**(3), 333–340 (2005)
14. Cardei, M., Cheng, M.X., Cheng, X., Du, D.-Z.: Connected domination in ad hoc wireless networks. In: Proceedings the Sixth International Conference on Computer Science and Informatics (CS&I'2002) (2002)

15. Cardei, M., Thai, M., Li, Y., Wu, W.: Energy-efficient target coverage in wireless sensor networks. IEEE INFOCOM, pp. 1976–1984 (2005)
16. Cardei, M., MacCallum, D., Cheng, X., Min, M., Jia, X., Li, D., Du, D.-Z.: Wireless sensor networks with energy efficient organization. J. Interconnect. Networks **3**(3–4), 213–229 (2002)
17. Chen, Y.P., Liestman, A.L.: Approximating minimum size weakly-connected dominating sets for clustering mobile ad hoc networks. In: Proceedings of the Third ACM International Symposium on Mobile ad hoc Networking and Computing, Lausanne, Switzerland (2002)
18. Cheng, M.X., Gong, X.: Maximum lifetime coverage preserving scheduling algorithms in sensor networks. J. Global Optim. **51**(3), 447–462 (2011)
19. Cheng, M.X., Ruan, L., Wu, W.: Achieving minimum coverage breach under bandwidth constraints in wireless sensor networks. INFOCOM, pp. 2638–2645 (2005)
20. Cheng, M.X., Ruan, L., Wu, W.: Coverage breach problems in bandwidth-constrained sensor networks. TOSN **3**(2), 12 (2007)
21. Cheng, X., Ding, M., Du, D.H., Jia, X.: Virtual backbone construction in multihop ad hoc wireless networks. Wireless Comm. Mobile Comput. **6**, 183–190 (2006)
22. Cheng, X., Huang, X., Li, D., Wu, W., Du, D.-Z.: A polynomial-time approximation scheme for minimum connected dominating set in ad hoc wireless networks. Networks **42**(2), 202–208 (2003)
23. Chvátal, V.: A greedy heuristic for the set-covering problem. Math. Oper. Res. **4**(3), 233–235 (1979)
24. Clark, B.N., Colbourn, C.J., Johnson, D.S.: Unit disk graphs. Disc. Math. **86**(1–3), 165–177 (1990)
25. Colbourn, C.J., Stewart, L.K.: Permutation graphs: connected domination and Steiner trees. Disc. Math. **86**(1–3), 179–189 (1990)
26. Dai, F., Wu, J.: An extended localized algorithm for connected dominating set formation in ad hoc wireless networks. IEEE Trans. Parallel Distrib. Syst. **15**(10), 908–920 (2004)
27. Dai, D., Yu, C.: A $(5+\varepsilon)$-approximation algorithm for minimum weighted dominating set in unit disk graph. Theor. Comput. Sci. **410**, 756–765 (2009)
28. Deering, S., Farinacci, D., Jacobson, V., Lui, C.-G., Wei, L.: An architecture for wide area multicast routing. In: Proceedings of ACM SIGCOMM 1994, pp. 126–135 (1994)
29. Ding, L., Wu, W., Willson, J.K., Du, H., Lee, W.: Construction of directional virtual backbones with minimum routing cost in wireless networks. In: 30th Annual Joint Conference of IEEE Communication and Computer Society (INFOCOM), pp. 1557–1565 (2011)
30. Ding, L., Wu, W., Willson, J.K., Du, H., Lee, W.: Efficient virtual backbone construction with routing cost constraint in wireless networks using directional antennas. IEEE Trans. Mobile Comput. **11**(7), 1102–1112 (2012)
31. Ding, L., Gao, X., Wu, W., Lee, W., Zhu, X., Du, D.-Z.: An exact algorithm for minimum CDS with shortest path constraint in wireless networks. Optim. Lett. (2010) published online
32. Ding, L., Gao, X., Wu, W., Lee, W., Zhu, Xu, Du, D.-Z.: Distributed construction of connected dominating sets with minimum routing cost in wireless network. In: Proceedings of the 30th International Conference on Distributed Computing Systems (ICDCS), pp. 448–457 (2010)
33. Ding, L., Wu, W., Willson, J.K., Du, H., Lee, W., Du, D.-Z.: Efficient algorithms for topology control problem with routing cost constraints in wireless networks. IEEE. Trans. Parallel. Distr. Syst. **22**(10), 1601–1609 (2011)
34. Ding, L., Wu, W., Willson, J.K., Wu, L., Lu, Z., Lee, W.: Constant-approximation for target coverage problem in wireless sensor networks. In: Proceedings of the 31st Annual Joint Conference of IEEE Communication and Computer Society (INFOCOM) (2012)
35. Douglas, R.J.: NP-completeness and degree restricted spanning trees. Disc. Math. **105**(1–3), 41–47 (1992)
36. Du, D.-Z., Ko, K.-I., Hu, X.: Design and Analysis of Approximation Algorithms. Springer, Berlin (2011)
37. Du, H., Pardalos, P.M., Wu, W., Wu, L.: Maximum lifetime connected coverage with two active-phase sensors. J. Global Optim. (2012), on line

References

38. Du, D.-Z., Thai, M.Y., Li, Y., Liu, D., Zhu, S.: Strongly connected dominating sets in wireless sensor networks with unidirectional links. APWeb, pp. 13–24 (2006)
39. Du, H., Wu, W., Ye, Q., Li, D., Lee, W., Xu, X.: CDS-based virtual backbone construction with guaranteed routing cost in wireless sensor networks. IEEE Trans. Parallel Distr. Syst., to appear
40. Du, D.-Z., Graham, R.L., Pardalos, P.M., Wan, P.J., Wu, W., Zhao, W.: Analysis of greedy approximation with nonsubmodular potential functions. In: Proceedings of SODA (2008)
41. Du, H., Ye, Q., Zhong, J., Wang, A., Lee, W., Park, H.: PTAS for minimum connected dominating set with routing cost constraint in wireless sensor networks. In: 4th Annual International Conference on Combinatorial Optimization and Applications (COCOA) (2010)
42. Du, H., Wu, W., Lee, W., Liu, Q., Zhang, Z., Du, D.-Z.: On minumum submodular cover with submodular cost. J. Global Optim. **50**(2), 229–234 (2011)
43. Du, H., Ye, Q., Wu, W., Lee, W., Li, D., Du, D.-Z., Howard, S.: Constant approximation for virtual backbone construction with guaranteed routing cost in wireless sensor networks. INFOCOM, pp. 1737–1744 (2011)
44. Duchet, P., Meyniel, H.: On Hadwiger's number and stability numbers. Annal. Disc. Math. **13**, 71–74 (1982)
45. Eriksson, H.: MBone: the multicast backbone. Commun. ACM **37**(8), pp. 54–60 (1994)
46. Erlebach, T., Mihalák, M.: A $(4+\varepsilon)$-approximation for the minimum-weight dominating set problem in unit disk graphs. WAOA, pp. 135–146 (2009)
47. Feige, U.: A threshold of $\ln n$ for approximating set cover. J. ACM **45**(4), 634–652 (1998)
48. Feige, U., Halldórsson, M.M., Kortsarz, G., Srinivasan, A.: Approximating the domatic number. SIAM J. Comput. **32**(1), 172–195 (2002)
49. Fodor, F.: The densest packing of 13 congruent circles in a circle. Beitrage Algebra Geom. **44**(2), 431–440 (2003)
50. Folkman, J.H., Graham, R.L.: A packing inequality for compact convex subsets of the plane. Canad. Math. Bull. **12**, 745–752 (1969)
51. Funke, S., Kesselman, A., Meyer, U., Segal, M.: A simple improved distributed algorithm for minimum CDS in unit disk graphs. ACM Trans. Sensor Net. **2**, 444–453 (2006)
52. Fujito, T.: Approximation algorithms for submodular set cover with application, IEICE Trans. Inf. Syst. **E83-D**(3), 480–487 (2000)
53. Gao, X., Huang, Y., Zhang, Z., Wu, W.: (6+epsilon)-approximation for minimum weight dominating set in unit disk graphs. COCOON, pp. 551–557 (2008)
54. Gao, X., Wang, Y., Li, X., Wu, W.: Analysis on theoretical bounds of approximating dominating set problems. Disc. Math. Algorithms Appl. **1**(1), 71–84 (2009)
55. Garg, N., Könemann, J.: Faster and simpler algorithms for multicommodity flows and other fractional packing problems. In: Proceedings of 39th Annual Symposium on the Foundations of Computer Science (FOCS), pp. 300–309 (1998)
56. Garey, M.R., Johnson, D.S.: The rectilinear Steiner tree problem is NP-complete. SIAM J. Appl. Math. **32**, 826–834 (1977)
57. Garey, M.R., Johnson, D.S.: Computers and Intractability: A Guide to the Theory of NP-Completeness. Freeman and Sons, San Francisco, CA (1979)
58. Gfeller, B., Vicari, E.: A faster distributed approximation scheme for the connected dominating set problem for growth-bounded graphs. In: Proceedings of the 6th Ad-Hoc, Mobile, and Wireless Networks International Conference (ADHOC-NOW 2007). Lecture Notes in Computer Science, vol. 4686, pp. 59–73. Springer, Berlin (2007)
59. Gibson, M., Pirwani, I.: Algorithms for dominating set in disk graphs: breaking the $\log n$ barrier. Algorithms–ESA, pp. 243–254 (2010)
60. Green, P.E.: Fiber-Optic Networks. Prentical-Hall, Cambrige, MA (1992)
61. Groemer, H.: Über die Einlagerung von Kreisen in einen konvexen Bereich. Math. Z. **73**, 285–294 (1960)
62. Guha, S., Khuller, S.: Approximation algorithms for connected dominating sets. Algorithmica **20**(4), 374–387 (1998)

63. Guha, S., Khuller, S.: Improved methods for approximating node weighted Steiner trees and connected dominating sets. Springer Lect. Notes Comput. Sci. **1530**, 54–66 (1998)
64. Hahn, R., Reichl, H.: Batteries and power supplies for wearable and ubiquitous computing. In: Proceedings of the 3rd International Symposium on Wearable Computers (1999)
65. Hoppe, R.: Bemerkungen de redaktion. Grunert Arch. Math. Phys. **56**, 307–312 (1874)
66. Huang, Y., Gao, X., Zhang, Z., Wu, W.: A better constant-factor approximation for weighted dominating set in unit disk graph. J. Combin. Optim. **18**(2), 179–194 (2009)
67. Kim, D., Wu, Y., Li, Y., Zou, F., Du, D.-Z.: Constructing minimum connected dominating sets with bounded diameters in wireless networks. IEEE Trans. Parallel Distrib. Syst. **20**(2), 147–157 (2009)
68. Kim, D., Zhang, Z., Li, X., Wang, W., Wu, W., Du, D.-Z.: A better approximation algorithm for computing connected dominating sets in unit ball graphs. IEEE Trans. Mobile Comput. **9**(8), 1108–1118 (2010)
69. Klein, P.N., Ravi, R.: A nearly best-possible approximation for node-weighted Steiner trees. J. Algorithms **19**, 104–115 (1995)
70. Kuhn, F., Zollinger, A.: Ad-hoc networks beyond unit disk graphs. In: Proceedings of the Joint Workshop of Foundation of Mobile Computing (DIALM-POMC) (2003)
71. Laskar, R., Pfaff, J.: Domination and irredundance in split graphs. Technical Report 430, Department of Mathematical Sciences, Clemson University (1983)
72. Li, M., Wan, P.J., Yao, F.F.: Tighter approximation bounds for minimum CDS in wireless ad hoc networks. ISAAC'2009. Lecture Notes in Computer Science, vol. 5878, pp. 699–709 (2009)
73. Li, D., Du, X., Hu, X., Jia, X.: Minimizing number of wavelengths in multicast routing trees in WDM networks. Networks **35**(4), 260–265 (2000)
74. Li, Y., Thai, M.T., Wang, F., Du, D.-Z.: On the construction of a strongly connected broadcast arborescence with bounded transmission delay. IEEE Trans. Mobile Comput. **5**(10), 1460–1470 (2006)
75. Li, Y., Thai, M.Y., Wang, F., Yi, C.-W., Wan, P.-J., Du, D.-Z.: On greedy construction of connected dominating sets in wireless networks. Wireless Commun. Mobile Comput. **5**, 927–932 (2005)
76. Li, D., Du, H., Wan, P.-J., Gao, X., Zhang, Z., Wu, W.: Minimum power strongly connected dominating sets in wireless networks. In: Proceedings of the 2008 International Conference on Wireless Networks (ICWN'08) (2008)
77. Li, D., Du, H., Wan, P.-J., Gao, X., Zhang, Z., Wu, W.: Construction of strongly connected dominating sets in asymmetric multihop wireless networks. Theoret. Comput. Sci. **410**(8–10), 661–669 (2009)
78. Liu, Q., Li, X., Wu, L., Du, H., Zhang, Z., Wu, W., Xu, Y.: A new proof for Zassenhaus–Groemer–Oler inequality. Disc. Math. Algorithms Appl. **4**(2), (2012) DOI: 10.1142/S1793830912500140
79. Min, M., Du, H., Jiao, X., Huang, X., Huang, S.C.-H., Wu, W.: Improving construction for connected dominating set with Steiner tree in wireless sensor networks. J. Global Optim. **35**, 111–119 (2006)
80. Moscibroda, T., Wattenhofer, R.: Maximizing the lifetime of dominating sets. In: Proceedings of the 5th IEEE International Workshop on Algorithms for Wireless, Mobile, Ad hoc and Sensor Networks (2005)
81. Mukherjee, B.: WDM-based local lightwave networks Part I: single-hop system. IEEE Networks **3**, 12–26 (1992)
82. Mukherjee, B.: WDM-based local lightwave networks Part II: multihop system. IEEE Networks **4**, 22–32 (1992)
83. Mustafa, N., Ray, S.: Improved results on geometric hitting set problems. Disc. Comput. Geometry **44**(4), 883–895 (2010)
84. Nemhauser, G.L., Wolsey, L.A.: Integer and Combinatorial Optimization. Wiley, New York (1999)
85. Oler, N.: An inequality in the geometry of numbers. Acta Math. **105**, 19–48 (1961)

References

86. Pandit, S., Pemmaraju, S.V., Varadarajan, K.R.: Approximation algorithms for domatic partitions of unit disk graphs. In: Dinur, L. et al. (eds.) APPROX and RANDOM 2009. Lecture Notes in Computer Science, vol. 5687, pp. 312–325 (2009)
87. Ramamurthy, B., Iness, J., Mukherjee, B.: Minimizing the number of optical amplifiers needed to support a multi-wavelength optical LAN/MAN. In: Proceedings of IEEE INFOCOM'97, pp. 261–268 (1997)
88. Ramaswami, R., Sasaki, G.: Multiwavelength optical networks with limited wavelength conversion. IEEE/ACM Trans. Networking **6**(6), 744–754 (1998)
89. Raz, R., Safra, S.: A sub-constant error-probability low-degree test, and a sub-constant error-probability PCP characterization of NP. In: Proceedings of 28th STOCS (1997)
90. Ruan, L., Du, D.-Z., Hu, X., Jia, X., Li, D.: Approximations for color-covering problems. In: Proceedings of 1st International Congress of Chinese Mathematicians, pp. 503–507. Beijing, China (1998)
91. Ruan, L., Du, D.-Z., Hu, X., Jia, X., Li, D., Sun, Z.: Converter placement supporting broadcast in WDM optical networks. IEEE Trans. Comput. **50**(7), 750–758 (2001)
92. Ruan, L., Du, H., Jia, X., Wu, W., Li, Y., Ko, K.-I.: A greedy approximation for minimum connected dominating set. Algorithmca, **329**(1–3), 325–330 (2004)
93. Salhieh, A., Weinmann, J., Kochha, M., Schwiebert, L.: Power efficient topologies for wireless sensor networks. ICPP'2001, pp. 156–163 (2001)
94. Sampathkumar, E., Walikar, H.B.: The connected domination number of a graph. J. Math. Phys. Sci. **13**(6), 607–613 (1979)
95. Sivakumar, R., Das, B., Bharghavan, V.: An improved spine-based infrastructure for routing in ad hoc networks. In: IEEE Symposium on Computer and Communications, Athens, Greece (1998)
96. Slijepcevic, S., Potkonjak, M.: Power efficient organization of wireless sensor networks. IEEE International Conference on Communications, pp. 472–476 (2001)
97. Solis-Oba, R.: 2-approximation algorithm for finding a spanning tree with maximum number of leaves. In: Proceedings of 6th European Symposium on Algorithms (ESA'98). Lecture Notes in Computer Science, vol. 1461, pp. 441–452. Springer, Berlin (1998)
98. Srinivasan, A., Wu, J.: TRACK: A novel connected dominating set based sink mobility model for WSNs. ICCCN, pp. 664–671 (2008)
99. Stojmenovic, I., Seddigh, M., Zunic, J.: Dominating sets and neighbor elimination based broadcasting algorithms in wireless networks. In: Proceedings of IEEE Hawaii International Conference on System Sciences (2001)
100. Sum, J., Wu, J., Ho, K.: Analysis on a localized pruning method for connected dominating sets. J. Inf. Sci. Eng. **23**(4), 1073–1086 (2007)
101. Thai, M.T., Du, D.-Z.: Connected dominating sets in disk graphs with bidirectional links. IEEE Commun. Lett. **10**(3), 138–140 (2006)
102. Thai, M.Y., Wang, F., Liu, D., Zhu, S., Du, D.-Z.: Connected dominating sets in wireless networks with different communication ranges. IEEE Trans. Mobile Comput. **6**(7), 721–730 (2007)
103. Vahdatpour, A., Dabiri, F., Moazeni, M., Sarrafzadeh, M.: Theoretical bound and practical analysis of connected dominating set in ad hoc and sensor networks. In: Proceedings of 22nd International Symposium on Distributed Computing (DISC), pp. 481–495 (2008)
104. Wan, P.J., Alzoubi, K.M., Frieder, O.: Distributed construction of connected dominating set in wireless ad hoc networks. ACM/Springer Mobile Network. Appl. **9**(2), 141–149 (2004). A preliminary version of this paper appeared in IEEE INFOCOM (2002)
105. Wan, P.-J., Alzoubi, K.M., Frieder, O.: A simple heuristic for minimum connected dominating set in graphs. Int. J. Found. Comput. Sci. **14**(2), 323–333 (2003)
106. Wan, P.J., Wang, L., Yao, F.F.: Two-phased approximation algorithms for minimum CDS in wireless ad hoc networks. IEEE ICDCS, pp. 337–344 (2008)
107. Wan, P.-J., Xu, X., Wang, Z.: Wireless coverage with disparate ranges. ACM Mobihoc (2011)

108. Wan, P.-J., Du, D.-Z., Pardalos, P.M., Wu, W.: Greedy approximations for minimum submodular cover with submodular cost. Comput. Optim. Appl. **45**(2), 463–474 (2010)
109. Wan, P.J., Huang, S.C.-H., Wang, L., Wan, Z., Jia, X.: Minimum-latency aggregation scheduling in multihop wireless networks. ACM MOBIHOC (2009)
110. Wan, P.-J., Wang, Z., Wan, Z., Huang, S.C.-H., Liu, H.: Minimum-latency schedulings for group communications in multi-channel multihop wireless networks. WASA (2009)
111. Wang, F., Thai, M.T., Du, D.-Z.: On the construction of 2-connected virtual backbone in wireless networks. IEEE Trans. Wireless Commun. **8**(3), 1230–1237 (2009)
112. Wang, L., Wan, P.-J., Yao, F.F.: Minimum CDS in multihop wireless networks with disparate communication ranges, WASA (2010)
113. White, K., Parber, M., Pulleyblank, W.: Steiner trees, connected domination and strongly chordal graphs. Networks **15**(1), 109–124 (1985)
114. Willson, J.K., Ding, L., Wu, W., Wu, L., Lu, Z., Lee, W.: A better constant-approximation for coverage problem in wireless sensor networks, preprint.
115. Willson, J.K., Gao, X., Qu, Z., Zhu, Y., Li, Y., Wu, W.: Efficient distributed algorithms for topology control problem with shortest path constraints. Disc. Math. Algorithms Appl. **1**, 437–461 (2009)
116. Wolsey, L.A.: An analysis of the greedy algorithm for submodular set covering problem. Combinatorica **2**(4), 385–393 (1982)
117. Wu, J., Li, H.: On calculating connected dominating set for efficient routing in ad hoc wireless networks. In: Proceedings of the 3rd ACM International Workshop on Discrete Algorithms and Methods for Mobile Computing and Communications, pp. 7–14 (1999)
118. Wu, J., Dai, F.: Virtual backbone construction in MANETs using adjustable transmission ranges. IEEE Trans. Mob. Comput. **5**(9), 1188–1200 (2006)
119. Wu, J., Gao, M., Stojmenovic, I.: On calculating power-aware connected dominating sets for efficient routing in ad hoc wireless networks. ICPP pp. 346–356 (2001)
120. Wu, J., Wu, B., Stojmenovic, I.: Power-aware broadcasting and activity scheduling in ad hoc wireless networks using connected dominating sets. Wireless Commun. Mobile Comput. **3**(4), 425–438 (2003)
121. Wu, J., Dai, F., Yang, S.: Iterative local solutions for connected dominating set in ad hoc wireless networks, MASS (2005)
122. Wu, J., Lou, W., Dai, F.: Extended multipoint relays to determine connected dominating sets in MANETs. IEEE Trans. Comput. **55**(3), 347–337 (2006)
123. Wu, W., Du, H., Jia, X., Li, Y., Huang, S.C.-H.: Minimum connected dominating sets and maximal independent sets in unit disk graphs, Theor. Comput. Sci. **352**(1–3), 1–7 (2006)
124. Xing, K., Cheng, W., Park, E.K., Rotenstreich, S.: Distributed connected dominating set construction in geometric k-disk graphs. IEEE ICDCS, pp. 673–680 (2008)
125. Zassenhaus, H.: Modern development in the geometry of numbers. Bull. Amer. Math. Soc. **67**, 427–439 (1961)
126. Zhang, Y., Li, W.: Modeling and energy consumption evaluation of a stochastic wireless sensor networks. preprint (2011)
127. Zhang, Z., Gao, X., Wu, W., Du, D.-Z.: A PTAS for minimum connected dominating set in 3-dimensional wireless sensor networks. J. Global Optim. **45**(3), 451–458 (2009)
128. Zhang, N., Shin, I., Zou, F., Wu, W., Thai, M.T.: Trade-off scheme for fault tolerant connected dominating sets on size and diameter. FOWANC, pp. 1–8 (2008)
129. Zhao, Y., Wu, J., Li, F., Lu, S.: VBS: Maximum lifetime sleep scheduling for wireless sensor networks using virtual backbones, INFOCOM pp. 366–370 (2010)
130. Zhong, C.: Sphere Packing. Springer, Berlin (1999)
131. Zhong, X., Wang, J., Hu, N.: Connected dominating set in 3-dimensional space for ad hoc network. In: Proceedings of IEEE Wireless Communications and Networking Conference (WCNC'07) (2007)

132. Zou, F., Li, X., Kim, D., Wu, W.: Construction of minimum connected dominating set in 3-dimensional wireless networks. In: Proceedings of Third International Conference on Wireless Algorithms, Systems and Applications (WASA 08) (2008)
133. Zou, F., Li, X., Gao, S., Wu, W.: Node-weighted Steiner tree approximation in unit disk graphs. J. Combin. Optim. **18**(4), 342–349 (2009)
134. Zou, F., Wang, Y., Xu, X., Du, H., Li, X., Wan, P., Wu, W.: New approximations for weighted dominating sets and connected dominating sets in unit disk graphs. Theoret. Comput. Sci. **412**(3), 198–208 (2011)

Index

Symbols
2-*independent number*, 187
H-free, 184
α-quasi unit disk graph, 70
$\partial \Omega$, 35
$ball_r(o)$, 64
k-Local Search, 155, 162, 165
k-independence number, 184
k-tight, 155, 162, 165
$sphere_r(o)$, 64
$disk_r(o)$, 35
(ROCα), 120
(ROC*α), 121
(ROC*0), 121
(ROC0), 119
(ROC1), 119
Voronoi cell, 152
Voronoi diagram, 154
Voronoi dual, 154
contraction of G, 184
contraction of an edge, 184
degenerate quadruple, 152
disk intersection graph (DIG), 151
geometrically redundant, 152
loose, 155
maximum gain, 146
minor of G, 184
move of L_i, 98
redundant, 155
shifted distance, 152
tight, 155
Gregory–Newton Problem, 63
configuration of sweep lines, 99
corner of the upper envelope, 98
regular polygon, 67
striding polygon, 67
3DDS, 7

CDS-SCHEDULING, 8
CONVERTER PLACEMENT, 5
LOCAL(e), 40, 127
MAX-LIFETIME CONNECTED-COVERAGE with two active phases, 107
MAX-LIFETIME COVERAGE, 106
MAX#CDS, 6
MAX#DS, 106
MIN-CDS, 2
MIN-SCDS, 13
MIN-SET-COVER, 13
MIN-SUBMODULAR-COVER, 18
MIN-WCDS, 17
MINW-BROADCAST, 29
MINW-CDS, 12
MINW-CSC with Two Active Phases, 108
MINW-CHROMATIC-DISK-COVER, 92
MINW-DS in Unit Disk Graphs, 77
MINW-DS on a Block B, 80
MINW-DS on a Cell e, 82
MINW-SENOR-COVER, 106
MINW-SENSOR-COVER, 113
MINW-SET-COVER, 12
NODE-WEIGHTED STEINER TREE, 12
PLANAR-4-CVC, 38
ST-MSP-IN-UDG, 45
SENSOR-COVER-PARTITION, 105
SENSOR-COVER-PARTITION with Separating Line, 113
SENSOR-COVER with Targets in Multi-Strips, 97
SENSOR-COVER with Targets in a Strip, 86
SET COVER, 3

A
active sensor set pair, 107
alive, 107

Index

B
black component, 47
boundary area, 39, 126
broadcasting tree, 29

C
CDS, 1
central area, 39, 127
circle$_r(v)$, 35
closed, 134
connected components, 69
connected domatic number, 6
connected dominating set, 1
connected domination number, 1
connected vertex cover, 38
controlled, 87
cover, 165, 170
coverage-preserving, 166

D
deficiency graph, 185
degenerate quadruple, 163
Disk-containment graphs, 133
domatic number, 6
dominating set, 1, 12
dominating tree, 176
domination-preserving, 157
double partition, 77

E
Euler characteristic, 58

F
face, 58
forbidden, 157
forbidden circle, 164

G
gain, 146
growth-bounded graph, 69

H
head, 29
heavy, 116
hit, 161
hitting set, 161

I
increasing, 17
independent, 47, 63

L
legal, 29
legal move, 99
light, 116
local independence number, 133
locality condition, 156, 162, 166
loose, 162, 165
lower area, 97
lower disk, 89
lower dumming disks, 97
lower envelope, 97

M
marginal value, 17
max leaf number, 1
minimal cover, 170
minimum CDS, 1

N
neighborhood area, 35

O
orphan, 29
out-arborescence, 29

P
partial CDS, 9
private, 171

R
redundant, 165
residue pair, 185
restricted, 187
restricted 2-independent set, 187
restricted connected 2-dominating set, 185
restricted perturbation, 163

S
sensor cover, 105
set cover, 3
simplex, 58
simplicial complex, 58
skyline, 114
spider, 29

Steiner nodes, 45
strongly connected dominating set (SCDS), 12
submodular, 17
sweep line, 98

T
terminals, 45
tight, 162, 165
topological control, 3

U
unit disk graph, 35
upper (dumming) disk, 97
upper area, 97
upper disk, 89
upper envelope, 97

W
weakly CDS (WCDS), 17

The manufacturer's authorised representative in the EU is Springer Nature Customer Service Centre GmbH, Europaplatz 3, 69115 Heidelberg, Germany. If you have any concerns regarding our products, please contact ProductSafety@springernature.com

Printed and bound by CPI Group (UK) Ltd, Croydon, CR0 4YY

23/03/2026

02076380-0002